# SLING BRAIDING
## TRADITIONS AND TECHNIQUES

From Peru, Bolivia, and Around the World

**RODRICK OWEN & TERRY NEWHOUSE FLYNN**

4880 Lower Valley Road · Atglen, PA 19310

**Other Schiffer Books by the Authors:**

*Andean Sling Braids: New Designs for Textile Artists,* Rodrick Owen and Terry Newhouse Flynn, ISBN 978-0-7643-5103-7

**Other Schiffer Books on Related Subjects:**

*Weaving Innovations from the Bateman Collection,* Robyn Spady, Nancy A. Tracy, Marjorie Fiddler, Foreword by Madelyn van der Hoogt, ISBN 978-0-7643-4991-1

*Leather Braiding,* Bruce Grant, ISBN 978-0-87033-039-1

*Handwoven Tape: Understanding and Weaving Early American and Contemporary Tape,* Susan Faulkner Weaver, ISBN 978-0-7643-5196-9

Copyright © 2017 by Rodrick Owen and Terry Newhouse Flynn

Library of Congress Control Number: 2017935612

All rights reserved. No part of this work may be reproduced or used in any form or by any means—graphic, electronic, or mechanical, including photocopying or information storage and retrieval systems—without written permission from the publisher.

The scanning, uploading, and distribution of this book or any part thereof via the Internet or any other means without the permission of the publisher is illegal and punishable by law. Please purchase only authorized editions and do not participate in or encourage the electronic piracy of copyrighted materials.

"Schiffer," "Schiffer Publishing, Ltd.," and the pen and inkwell logo are registered trademarks of Schiffer Publishing, Ltd.

Designed by Brenda McCallum
Cover design by Brenda McCallum
All photos were taken by and all diagrams were created by the authors unless otherwise noted.
Type set in Futura/Times

ISBN: 978-0-7643-5430-4

Printed in China

Published by Schiffer Publishing, Ltd.
4880 Lower Valley Road
Atglen, PA 19310
Phone: (610) 593-1777; Fax: (610) 593-2002
E-mail: Info@schifferbooks.com
Web: www.schifferbooks.com

For our complete selection of fine books on this and related subjects, please visit our website at www.schifferbooks.com. You may also write for a free catalog.

Schiffer Publishing's titles are available at special discounts for bulk purchases for sales promotions or premiums. Special editions, including personalized covers, corporate imprints, and excerpts, can be created in large quantities for special needs. For more information, contact the publisher.

We are always looking for people to write books on new and related subjects. If you have an idea for a book, please contact us at proposals@schifferbooks.com.

This book is dedicated to the ancient and contemporary
Andean sling braiders who were the originators of these unique patterns
as we know them today.

# CONTENTS

*Denotes braids that can be made on the braiding stand without a core stand.

| | |
|---|---|
| Preface | 8 |
| Acknowledgments | 10 |

### 1. A History of the Sling — 11

**Early History of the Sling in the Non-Andean World** — 11
- The Middle East and the Mediterranean — 12
- *Egypt* — 13
- The Roman Empire — 13
- Iron Age Britain to Anglo/Norman Britain — 14
- Nepal, Tibet, and Western China — 15
- The Pacific Islands — 17
- North America — 18

**A Selection of Contemporary Slings from Around the World** — 18
- Saudi Arabia — 18
- The Pitt Rivers Museum Collection — 19

**Slings in Pre-Hispanic Andean Cultures** — 21
- *Huaca Prieta, Asia, and Otuma (3200–1200 BCE)* — 23
- *The Paracas and Nasca Cultures* — 24
- *Ancón* — 26
- *Pachacamac* — 27
- *The Moche Empire* — 27
- *Tiwanaku and Wari* — 28
- *The Inca* — 29
- Historic Documentation on Military Use of the Sling — 30
- Inca Maturity Rites for Boys — 30
- Inca Agricultural and Hunting Practices — 31

**Andean Slings After the Fall of the Inca Empire** — 31

**Contemporary Andean Slings** — 31

### 2. Braiding Equipment — 35

- **The Core Braiding Frame** — 35
- **Bobbins** — 35
- **Counterbalance Bag** — 36
- **Warping Equipment** — 36
- **Accessories** — 37

| | |
|---|---|
| **Counterbalance Weights** | 38 |
| **The Braiding Card** | 38 |

## 3. Core Braiding Yarns: History, Contemporary Choices, and Design Considerations ........ 39

| | |
|---|---|
| **Pre-Hispanic Fibers, Spinning, and Dyes** | 39 |
| Pre-Hispanic Camelid Breeds | 39 |
| Pre-Hispanic Dyes | 41 |
| **Contemporary Andean Yarns for Slings** | 42 |
| **Contemporary Andean Yarns for Festival Braids** | 42 |
| An Ancient and Contemporary Yarn Comparison | 42 |
| **Yarn and Fiber Characteristics** | 43 |
| Yarn Construction Basics | 43 |
| *Western handspun yarns.* | 43 |
| *Andean handspun yarns.* | 43 |
| *Manufactured yarns.* | 43 |
| **Fibers for Braiding** | 44 |
| Wool | 44 |
| Alpaca and Llama Yarns | 46 |
| Cotton | 48 |
| Other Fibers | 48 |

## 4. Preparing the Warp, Setting Up for Braiding, and Basic Working Methods ........ 49

| | |
|---|---|
| **Planning the Braid Diameter** | 49 |
| **Planning the Warp Length** | 49 |
| **Winding a Warp for Starting with a Tassel** | 49 |
| Method 1: Warping for One End per Bobbin | 50 |
| Method 2: Warping for Multiple Ends per Bobbin | 51 |
| Binding the Tassel | 51 |
| **Winding Longer Warps** | 52 |
| **Method A for Beginning with a Blunt End: Using a Larkshead Knot** | 52 |
| **Setting Up and Working on the Card** | 53 |
| Tensioning on the card | 53 |
| **Setting Up and Working on the Stand** | 53 |

| | |
|---|---|
| *Attaching bobbins to the warp.* | 53 |
| *Making a weaver's knot.* | 53 |
| *Making the slipping hitch.* | 54 |
| *Bobbin working height.* | 54 |
| **Setting Up a Three-color Core Braid Sample** | 55 |
| **Choosing Counterbalance Weights** | 57 |
| Counterbalance Weights | 57 |
| **Attaching the Counterbalance Bag** | 57 |
| **Developing a Comfortable Working Position** | 58 |

## 5. Braid Designs 8.1–32.1 ........ 59

| | |
|---|---|
| **Eight-strand Braid Designs 8.1 and 8.2** | 59 |
| Interpreting Braiding Diagrams | 59 |
| *Reading rotation diagrams.* | 59 |
| *Reading diagrams for the stand.* | 60 |
| *Reading diagrams for the card.* | 61 |
| Finishing with a Tassel | 61 |
| *Design 8.1 Spiral Braid with Variations | 62 |
| *Design 8.2 Round Braid | 62 |
| Sixteen-strand Braid Designs 16.1–16.3 | 63 |
| *Design 16.1 Spiral Braids | 63 |
| *Design 16.2. Square Braid Variations | 64 |
| *Design 16.3 Diamonds and Zigzags | 64 |
| *Key moves for making inversion steps on the card.* | 65 |
| **Chevron Designs 24.1–24.3** | 66 |
| Making 24-strand Chevron Patterns | 67 |
| Two Methods of Making 24-strand Square Braids on the Core Stand | 67 |
| Working Tension for Card and Stand | 67 |
| *Tensioning for the card.* | 67 |
| *Tensioning for stand braiding.* | 68 |
| *Counterbalance weight.* | 68 |
| Checking for Errors | 68 |
| *Using a lifeline.* | 68 |
| *Design 24.1. Classic Chevron Braid in Two-, Three-, and Four-color | 69 |

| | |
|---|---|
| Design 24.2. Three-color Chevrons with Core Changes | 69 |
| Color Substitution when Sampling | 69 |
| Interpreting Core Exchange Diagrams | 69 |
| Steps for Exchanging the Core on the Card | 70 |
| Adjusting the Tension for Card Braids | 71 |
| *Design 24.3. Chevron Braid with Vertical Stripes | 71 |

**Designs *24.4 and 24.5: Making Diamond and Chevron Designs Using Inversion Moves** ............ 72

| | |
|---|---|
| Creating Patterns with Inversion Moves: Designs 24.5–24.13 | 72 |
| Walk-through of Making Single Diamond Pattern Design 24.6.1 | 73 |
| Adjusting Tension for the Stand and the Card after Inversion Steps (or Mind the Gap) | 75 |

**Designs 24.6–24.10. Two-color Diamond Variations** ............ 76

**Design 24.11. Alternate Method for Creating Diamonds** ............ 77

**Multicolor Diamond Designs 24.12–32.1** ............ 78

| | |
|---|---|
| Design 24.12. Braid Architecture Explained: Changing between Diamond and Chevron Patterns | 78 |
| Design 24.13. Embossed Diamonds with Thick and Thin Yarns | 79 |
| Design 24.14. Four-color Diamonds with Cores of 16 Strands | 81 |
| *Design 24.15 Three-color Diamonds Made without Cores | 81 |
| *Design 32.1 Textured Four-color Diamonds Made without Cores | 82 |

**The Designs** ............ 83

## 6. Beginnings, Endings, and Embellishments ............ 133

**Braided Loops** ............ 133

| | |
|---|---|
| Starting a Braid with an 8-strand Loop on the Stand | 133 |
| *Preparing the warp and setting up for a loop.* | 134 |
| *Making the 8-strand braid for the loop.* | 136 |
| Starting an 8-strand Loop on the Card | 137 |
| Making a Split Loop | 139 |

**Starting a 16-strand Braid with a Blunt End, Method B** ............ 143

| | |
|---|---|
| Attaching Bobbins to Both Ends of a Warp | 143 |
| Making a Sample of Design 16.1.3 with a Blunt-end Beginning | 144 |

**Starting a 24-strand Braid with a Blunt End** ............ 146

**Starting a 24-strand Braid with a Hollow-braid Wrapping** ............ 146

| | |
|---|---|
| Setting Up a Hollow Braid | 147 |
| Working a Hollow Braid over a Core | 147 |
| Transitioning to a Chevron or Diamond Pattern | 148 |

**Stitched Finishes** ............ 149

**Stitched "Beads"** ............ 149

| | |
|---|---|
| Inserted Tassels with Two Styles of Binding | 150 |
| *Inserting a tassel skirt.* | 152 |
| *Tassel binding method 1.* | 152 |
| *Tassel binding method 2.* | 152 |

**Working Braids over Sculptural Forms** ............ 153

**Cross-Knit Looping** ............ 153

| | |
|---|---|
| Using Tubular Cross-knit Looping to Finish or Embellish a Braid | 154 |

**Finishes for Sling Cradles** ............ 155

| | |
|---|---|
| Cross-knit Looping Used for Finishing Sling Cradle Selvages | 155 |

## 7. Making an Andean-style Sling ............ 157

**Traditional Macusani and Ilave Sling Making Methods** ............ 157

**Calculating the Warp Length for a Sample Sling** ............ 160

**Prepare Four Warp Sections** ............ 160

**Making the Finger Loop and First Sling Cord** ............ 161

Making the 8-strand Finger Loop ............ 161

| | |
|---|---|
| Transitioning from 16 Strands to 24 Strands | 161 |
| Adding 12 White Threads | 162 |
| Braiding the 24-strand Diamond Pattern | 164 |

**Preparing the Warp for Weaving the Cradle** ............ 165

| | |
|---|---|
| Binding the Thread Groups | 167 |
| Organizing Warp Groups by Twining | 167 |

**Weaving the Cradle** .......................................... 168
    Tapestry Weaving Basics ............................. 168
    Cradle Design ............................................. 169
    Cradle Gallery ............................................. 170
    Widening the Warp Before Weaving
    the Pattern .................................................. 170
    Setting Up the Warp for Weaving ............... 172
    Weaving the Pattern and Creating the Slit ......... 172
    *Cradle design 1.* ......................................... 172
    *Cradle design 2.* ......................................... 172
    Dividing for the Slit and Weaving
    the Triangles. .............................................. 172
    Outlining and Filling the Diamond ............. 173
    Closing the Slit and Tapering the Cradle ........ 173

**Making the Second Braid** .............................. 173
    Setting Up the Stand for Making
    the Second Braid ........................................ 173
    Adding Warp D and Braiding the
    Diamond Pattern ........................................ 174
    Completing the Diamond Pattern ................ 176
    Reducing the Braid from 24 Strands
    to 16 Strands ............................................... 176

**Additional Sling Construction Methods** ........ 180
    Laverne Waddington's Story ........................ 180
    Ben Turner's Story ...................................... 181

**Appendix A. Plans for a Core Frame** ............... 183
**Appendix B. Plans for Making a
32-slot Card for Core Braiding** ......................... 187
**Appendix C. Rotational Move Diagrams
for 24-strand Core Braids, Methods 1 and 2** ........ 190
**Appendix D. Design Planning Diagram** .......... 194
**Appendix E. Reading Core
Exchange Diagrams** ........................................ 195
**Appendix F. Comparison of Sling
Dimensions** ..................................................... 202

Notes .................................................................. 203
Bibliography ...................................................... 205

# PREFACE

When we submitted our first manuscript to Schiffer Publishing in May of 2015, it never crossed our mind that we would be writing two books on Andean braids instead of one. We had, while working on the initial manuscript, become aware that it was growing a bit . . . large, despite leaving out many things of interest to us and to potential readers.

In the fall of 2015 we a received a phone call from our editor, Sandra Korinchak, with the news that the manuscript was indeed large. They could go ahead and publish it if we were comfortable with it being an expensive volume and thus less accessible to a wider audience. Or, was it possible to divide the manuscript into two books, each with its own focus?

After some discussion, we saw that this could indeed work. The first book, *Andean Sling Braids: New Designs for Textile Artists*, had the craftsperson as its focus. Here at last is book two with its focus on slings within their historical and cultural contexts. We have written it for both the craftsperson interested in making these unusual braids, and for an audience interested in the history of weapons, their use, and the practical aspects of making slings.

The world over, slings are made up of the same basic parts: two sling cords on either side of a cradle (or pouch) that is designed to hold a stone or projectile. Chapter 1, *A History of the Sling* explores the similarities and differences between the sling in Andean and non-Andean cultures. In some geographic areas, particularly in Asia, Polynesia, and the Andes, slings are constructed using a variety of interesting and often uncommon textile techniques using braiding, twining, ply-split braiding, knotting . . . . We have credited our sources and hope that the bibliography will help you in your own research whether it is for learning about ancient cultures, for making slings, for making jewelry, or something else.

The next three chapters cover essential knowledge regarding braiding equipment; recommended yarns (with historical information on Andean fibers and yarn); how to wind a warp, set up a braiding card or Japanese-style braiding stand with bobbins, and use the core frame, which allows core threads to be suspended above a braiding stand.

Chapter 5 contains the braid designs, using from 8 to 32 strands, with and without cores. Andean core braids are architectural marvels. The book starts with explaining a few simple 8- and 16- strand braids. Then Design 24.1, a 24-strand square braid is introduced. This is the classic core braid structure with six threads on each of the four cardinal points of a braiding stand or card. The north and south threads interlace, then the east and west threads. A core of threads can be added to the center of the braid from which colors can be substituted. This is where the thread games begin!

For novice braid makers and sling makers, we have included some of the essential designs found in our first book. We have

included instructions for making core changes and some information on braid structure. (If you want to delve deeper into Andean braid patterns and braid structure, our first book is a good source for additional braid designs, from simple to complex.) We have included many new designs in this book.

In our first book, we took the classic approach of detailing each set of steps necessary to make each braid design. (This can take many pages for a braid that uses a simple strategy but requires many steps to get back to the initial color order.) We couldn't do this and have room for the designs that didn't make it into book one. Neither did we want novice braiders, who did not own the first book, to lack enough of the classic designs. The solution is found in Table 5.1 (on p. 121) that shows which sequences of steps are related to each pattern's architecture. As you work with the chart, you will hopefully see how easy it is to change from one design to another, i.e., from red and white diamonds to red and black diamonds with a black outline. We hope you will be inspired to use this tool to create your own designs.

Chapter 6, *Beginnings, Endings, and Embellishments*, will be of interest to both those interested in Andean textiles and Kumihimo enthusiasts. We have shared both traditional Andean finishes and some of our own favorites, things you might learn from us in a class but for which we had never gotten around to writing specific diagrams and instructions. These are the techniques that will help you use braids effectively in creative applications.

The book culminates with Chapter 7, *Making an Andean-style Sling* with step-by-step instructions that detail how to do this. The basic process entails making an 8-strand braid in the middle of a warp; joining the two ends to create a finger loop; making the first 16-strand braided sling cord; adding 8 threads to change it to a 24-strand structure; finally adding a further 12 threads to make 36 strands for braiding a different two-color diamond; before transitioning to weaving the cradle; then reversing the process to mirror the first sling cord. We have included many photographs of slings throughout the book. In addition, Appendix F features a table with data on the slings in Rodrick's collection.

We chose not to describe braiding in the hand (or fist) as this technique is well covered in *Sling Braiding of the Andes* by Adele Cahlander, with Elayne Zorn and Ann Pollard Rowe. (See bibliography.)

Our research journey has convinced us, at least in regard to sling braiding, that some ancient Andean braiding traditions are still carried on. We hope that our book will lead to a greater appreciation of Andean braided textiles. We hope this book inspires many braiding adventures!

# ACKNOWLEDGMENTS

So many people have helped us with this project. Our sincere thanks go out to each and every one of you.

*In the book's infancy:*

To past and present staff at Schiffer Publishing: Stephanie Daugherty, for suggesting that we talk to Schiffer about the book we had started writing on Andean sling braids. To Pete and Nancy Schiffer for your interest in our project and for giving us the opportunity to share our interest in Andean braided textiles.

To our design testers Suzi Siorek and Tim Clark who spent many hours testing the braiding instructions, and offering valuable (and sometimes humorous) feedback.

*In the book's long adolescence while we refined the writing, graphics, and illustrations:*

To our proofreaders Anne Rock, Richard Sutherland, and Duane Wakeman for your many insightful questions regarding this second book.

To Richard Sutherland for kindly taking on the task of formatting the bibliography and endnotes.

To novice sling makers Doug Newhouse, Anne Rock, and Valerie Fry, for testing our sling making instructions and for successfully completing their slings.

To Laverne Waddington and Ben Turner for sharing stories about their sling making experiences.

To Dr. Barbara Wolfe, anthropologist, for advising us on the complex history of the Andes and for making recommendations on the history chapter. Thanks for putting us in touch with many people in the field of Andean studies. To the non-profit organization Ayni, where we first found Eleuteria Nuñez Huamani's inspiring sling braids. Thanks to master weavers Oscar Huarancca Gutierrez, Alfredo Jayo Rojas, and Fermin Aybar Ayala for answering questions about sling making.

A picture is worth a thousand words! Many thanks to the following people who generously shared photographs from their travels or took them specifically for us: Hedy Hollyfield, Carole Thorpe, Gina Corrigan, Joy Totah Hilden, Ingrid Crickmore, and Danielle Murphy. Many thanks go to Andy Owen for the many line drawings that appear in the book. Thanks to Kelly Newhouse for lending your hands to the card braiding photos.

Our book was made richer by the assistance of many museums and individuals with answering questions and allowing us to use or purchase images of their sling collections, including: Donald Proulx, professor emeritus of the University of Massachusetts for helping us find slings decorating ceramics; Thom Richardson, Keeper of Armour and Oriental Collections at the Royal Armouries Museum, for giving permission to use material from his published work; Christopher Philipp, collections manager, The Field Museum in Chicago; the Pitt Rivers Museum for allowing the photographing of part of their sling collection.

To Chris Harrison who shares diverse sources of slinging information on his website slinging.org.

*In its final stage of production:*

Special thanks to our editor, Sandra Korinchak, for your enthusiasm and ability to explain the intricacies of fine-tuning a manuscript! Thanks also to our book designer Brenda McCallum, whose skill not only makes this volume beautiful, but makes it easier for the reader to use.

# 1. A HISTORY OF THE SLING

**FIGURE 1.1.**
Contemporary Andean sling made with alpaca yarn.
*Collection of R. Owen.*

Most slings consist of equal lengths of cord on either side of a wider section, called a cradle (or pouch), which is designed to hold a missile, usually a stone. (See figure 1.1.) The cradle is the focal point of a sling. Its function is to hold the missile securely until it is released and sent to its target. A finger or wrist loop is often worked into the end of one or both sling cords to keep the sling in the hand once the missile is thrown. When used for herding, the sling is often twirled in the air or snapped like a whip. Festival slings, which are more decorative and less functional, may have smaller cradles. (See figure 1.39.)

## Early History of the Sling in the Non-Andean World

The long-range weapons of sling and bow appeared during the Mesolithic period, 12,000–8,000 BCE, along with combat weapons—the short sword and mace.[1] The sling was easy to make from such readily available materials as sinew, plant fibers, animal hide, and hair.[2] Initially, slings were used to hunt game and to ward off predators or outsiders. Their use would have been limited to areas where loose stones were widely

**FIGURE 1.2.**
Worldwide distribution of slings used as weapons from prehistory to recent times (1973). *Courtesy of Scientific American. All rights reserved.*

available.[3] Figure 1.2 shows the worldwide distribution of sling weapon use from prehistory to the 1970s from an article by Manfred Korfmann. If his survey had included slings used for herding, the distribution would be considerably larger.

Later, when armies incorporated slingers as part of their troops, slingshot was crafted from clay, limestone, chalk, and lead. Lead *plummets* (from the Old French for "ball of lead") or *glandes* (a form between acorns and almonds) were cast in molds. The lead shot pictured in figure 1.3 was found at St Albans, Hertfordshire, UK, and is believed to date to the middle of the first century.[4] Slingshot in various cultures was stamped or incised with text or symbols. Writing about the Cretan tradition of inscribing shot, archaeologist Amanda Kelly says, "Slingshots bearing text are illuminating artefacts as not only can they reflect military action, leadership, civic affiliation and ethnicity, but they can also occasionally offer an insight into the psyche of their associated military personnel."[5] Shot from Roman, Greek, and other cultures often carried inscriptions, such as "An Achaean blow," "Your heart for Cerberus," or (sometimes humorously), "Ouch."[6]

While some slingshot is difficult to distinguish from common stones, the shaped and finely polished river stones from the Pacific islands of New Caledonia are well crafted. Figure 1.4 shows stones similar to ones collected during Captain James Cooke's exploratory voyages in the mid- to late 1700s.

**FIGURE 1.4.**
Shaped river stones and pouch from New Caledonia.
*Pitt Rivers Museum, University of Oxford, 1903.54.11.*

**FIGURE 1.3.**
Molded lead shot found at Windridge Farm, St. Albans, Hertfordshire, UK, thought to date to the mid-first century.
*Courtesy of St Alban's Museum SL1556.1.*

## The Middle East and the Mediterranean

Ancient Greek writings include a few references to slings (*funda*):

••••••••••••••••••••••••••••••••••••••••••••••••••••

The light troops of the Greek and Roman armies consisted in great part of slingers (*funditores*, σφενδονῆται). In the earliest times, however, the sling appears not to have been used by the Greeks. But in the times of the Persian wars slingers had come into use; for among the other troops which Gelon offered to send to the assistance of the Greeks against Xerxes, mention is made of 2000 slingers. At the same time, it must be stated that we rarely read of slingers in these wars. Among the Greeks, the Acarnanians in early times attained the greatest expertness in the use of this weapon (Thuc. II.81); and at a later time the Achaeans were celebrated as expert slingers.[7]

••••••••••••••••••••••••••••••••••••••••••••••••••••

The slingers who enjoyed the greatest celebrity were the natives of the Balearic Islands in the Mediterranean Sea. Balearic slings were (and still are) made from plant fiber. As slingers, the Balearians served as mercenaries, first under the Carthaginians and afterwards under the Romans.[8] The tale of the role Balearic mothers played in their sons' slinging expertise is recounted in many places and is a story still heard on the islands: "The inhabitants of the Balearic Islands are said to have been the inventors of slings, and to have managed them with surprising dexterity, owing to the manner of bringing up their children. The children were not allowed to have their food by their mothers till they had first struck it with their sling [stone]."[9]

*Egypt*

Egyptian tomb paintings show slings used for hunting. Two of the oldest slings yet discovered were found in the tomb of Tutankhamen who ruled ca. 1332–1323 BCE. (Andean slings predate these.) These finely plaited slings were likely intended for the departed pharaoh to use in hunting game.

The linen sling in figure 1.5, found at El-Lahun, Faiyum, Egypt, dates to the Twenty-second Dynasty (800 BCE). It features a finger loop made from a flat braid, finely braided cords (one is a fragment), and a diamond-shaped woven cradle. Thom Richardson noted in his analysis that the structure of the sling cords is almost identical to braids found on nineteenth- and twentieth-century Andean slings.[10]

FIGURE 1.6.
Roman slingers. From *Greek and Roman Antiquities*, John Murray, London, 1875. Re-drawn by Andy Owen.

## The Roman Empire

In the Roman Empire, slingers were an integral part of the army. The fourth century writer Vegetius listed the training necessary to make a good soldier in his book, *De Re Militari*:

Recruits are to be taught the art of throwing stones both with the hand and sling . . . Soldiers, notwithstanding their defensive armor, are often more annoyed by the round stones from the sling than by all the arrows of the enemy. Stones kill without mangling the body, and the contusion is mortal without loss of blood. It is universally known the ancients employed slingers in all their engagements. There is the greater reason for instructing all troops, without exception, in this exercise, as the sling cannot be reckoned any encumbrance, and often is of the greatest service, especially when they are obliged to engage in stony places, to defend a mountain or an eminence, or to repulse an enemy at the attack of a castle or city.[11]

FIGURE 1.5.
A reproduction of a linen sling *(right)*, and the fragment that was found at El-Lahun, Egypt, from the Late Third Intermediate Period (about 800 BCE). Copyright: Petrie Museum of Egyptian Archaeology, University College, London. UC6921.

Figure 1.6 is an illustration of two slingers from a carving on Trajan's Column. The freestanding column with its spiral bas relief commemorates the Roman emperor Trajan's victory in the Dacian Wars and dates to 113 CE.[12]

## Iron Age Britain to Anglo/Norman Britain

Iron Age hill-forts with caches of slingstones and molded clay projectiles dating to 400–150 BCE have been found across Britain. Examples from several sites include 22,600 pebbles recovered from Maiden Castle and 11,000 from Danebury. They were made of three types of material: natural water worn pebbles (sea-rounded, riverine, or glacially derived); baked and unbaked clay ovoids; and occasionally, carved stone, usually chalk, that weighed between 20 and 30 grams.[13] " . . . British Iron Age hillforts were specifically designed to be defended against slingstone barrages and in turn facilitated the use of slings and hand hurled stones by defenders."[14]

During the Anglo-Saxon period (410–1066 CE), slings were in common use by the military and occasionally appear in illustrations. Figure 1.7 is a reproduction of a drawing in the Cotton Collection that dates from the eighth century. On the right, staff-slings (called *fustibalus* by the Romans) were used for besieging cities and in naval engagements.[15]

At the battle of Hastings, " . . . slingers formed part of both armies: from this period until the close of the fourteenth century they formed an important element in every military expedition."[16] Additional evidence that slingers played a role in military campaigns during the Anglo/Norman period is found in *Chronicon ex Chronicis*. Just as King Harold thought his enemies were subdued, he received news that William, later called the Conqueror, "had arrived with a countless host of horsemen, slingers, archers, and foot-soldiers, having brought with him powerful auxiliaries from all parts of Gaul, and that he had landed at a place called Pefnesea [Pevensey]."[17] So slingers played a role in one of England's most significant battles.

FIGURE 1.7.
Engraving showing Anglo-Saxon slinging styles by Joseph Strutt based on old manuscripts.

## Nepal, Tibet, and Western China

Nepalese slings, once used as weapons, are still used to herd sheep and scare away predators. They also had ritual significance associated with boys' maturation traditions. When a boy reached the age of seven, he received his first sling. Over the course of the next seven years, he was given a new sling with a special name at each birthday:

> Small sling of wool, joy of the child
> sling of many colors
> sling of the shepherd
> sling of the eight strands
> sling of the warrior
> sling of nine eyes
> sling of the magician[18]

FIGURE 1.8.
*Tibetan shepherd braiding a sling.*
*Courtesy of Gina Corrigan.*

FIGURE 1.9.
*Tibetan Woman Using a Sling for Throwing Stones* by Arnold Henry Savage Landor (1905).

The *nine-eyed sling* is traditionally used by lamas (priests) during festivals. It has a cradle with a pattern of nine black and white stripes woven from animal hair. The evil spirits are first trapped within a *torma,* a conical-shaped ritual object made mostly of barley flour and butter. The lama uses the sling to cast the torma away from the village, thus ridding it of the evil spirits.[19]

Sling-makers in Tibet, Western China, and Nepal frequently use multiple textile techniques in a single sling as do sling-makers in the Andes. The braided portions of Tibetan and Nepalese slings use 16-strand spiral, chevron, or diamond patterns. The Tibetan braider shown in figure 1.8 is making a 16-strand spiral braid. Extra threads are added in order to make the transition to the cradle, which is worked in ply-split braiding.

Figure 1.9 is an illustration from *Tibet & Nepal Painted and Described* by A. Henry Savage Landor, 1905, in which he depicts a Tibetan woman using a sling. Her sling features a wrist loop, as do some contemporary slings from Tibet.

The Tibetan herding sling in figure 1.10 features both a braided loop and a split loop on the same cord. The cradle is twined. The ceremonial sling shown in figure 1.11 was collected by Mary Knipple in the southwestern part of the Autonomous Tibetan Region. It uses a tufting technique worked over a braided and ply-split cord. (Tufting or *furring* was also used in some pre-Hispanic Andean slings. Raoul d'Harcourt described three possible ways of making it.[20]) Natural colors are common in working slings while black, white, red, and bright colors are frequently used in ceremonial slings.

**FIGURE 1.10.**
A well-worn herding sling from Tibet with a twined cradle.
*Collection of R. Owen.*

**FIGURE 1.11.**
Ceremonial sling from Tibet.
The structure supporting the tufting is ply-split braiding.
*Collection of Mary Knipple.*

The colorful cotton sling shown in figure 1.12 and the black, white, and red sling shown in in figure 1.13 were collected between 1990 and 2001 from Qinghai Province, located in western China near the Autonomous Tibetan Region. They use a variety of textile techniques including braiding, weaving, ply-split braiding, and wrapping.

FIGURE 1.12.
Bright cotton sling using multiple textile techniques, from Yushu, Qinghai Province, China. *Courtesy of Gina Corrigan.*

FIGURE 1.13.
Sling from Qinghai Province, China, made of sheep's wool or yak fiber. *Courtesy of Gina Corrigan.*

## The Pacific Islands

Slings were a common weapon in the Pacific Islands and were mentioned in the writings from early European expeditions like those of Captain James Cooke. Slings were made of plant fibers; some were constructed using interesting textile techniques, including plaiting. Few slings survived due to the humid conditions but a variety of finely made slingstones are plentiful. Collections from early explorers can be found in the Pitt Rivers Museum, the Bernice Pauahi Bishop Museum, as well as others.

On the island of Ponape, Federated States of Micronesia, the sling was the favored weapon until the introduction of firearms by New England whalers. The slings were "plaited out of strips of Kalau or Hibiscus bark, or else out of the Cinnet fibre, or that of the Nin tree bark."[21] (Kalau, also known as *Hau* or sea hibiscus, comes from a small evergreen tree while cinnet is another name for coconut fiber.)

Slings were the preferred projectile weapons of Polynesia. They were made of coir (coconut fiber). Like Andean, Tibetan, and Nepalese slings, they used a variety of textile techniques. Peter Henry Buck described a partially decayed sling that dates from about 1819:

The edges of the pouch were formed by single lengths of sennit, but the body consisted of coir two-ply twist, the whole being kept together by crossing rows of *oronga* thread in what seemed to resemble the New Zealand two-pair interlocking weft used in weaving. At the commencement of the pouch, the coir cords were included in the *oronga* crossing weft strokes in pairs for a length of 57 mm. and then the cords were separated to be caught in the weft singly to widen the middle part of the pouch . . . The paired cords at the pouch ends were plaited in a six-ply braid to form the strings of the sling. One string without a loop was 1,000 mm. long. The other string was plaited as a six-ply braid for 910 mm. and then divided into two equal sets, which were finished off as three-ply braids. The ends were knotted together to form a loop 70 mm. long. The string was drawn back through the end loop to form a larger loop that would encircle the right wrist.[22]

In Mariana, slingstones are primarily associated with the late period of their prehistory, beginning some 800–1000 years ago. This marked proliferation of slingstones may have signaled the advent of warfare.[23] An account by Father Pedro Coomans in *The History of the Mariana Island Mission for the 1667-1673 Period* described sling use similar to that experienced by conquistadors in Peru:

> Their offensive weapons include the sling, which they aim very skillfully at the head . . . . They whirl and shoot those so violently. Should it make an impact upon a more delicate part, like the heart, or the head, the man is flattened on the spot. Then, if envy would make them want to burn a house from a distance, they would stuff the perforated side of it (the slingstone) with tow burning with a very ferocious fire, which, with a swift movement became a flame, and sail away to seek shelter in enemy houses.[24]

## North America

Few North American slings from pre-European contact have survived. Wide use of the sling for hunting birds and small game is not questioned, but there is disagreement as to the use of the sling in warfare. Collected evidence is scarce and stories sometimes conflict.[25]

A sling collected in 1912 at Lovelock Cave, Nevada, is rare due to its age. Carbon dating suggested a date of 792–272 BCE. It is made of vegetable fiber, possibly dogbane or Indian hemp. It was "wrapped with a compound necklace of Olivella-shell beads and was found about the neck of a partially mummified child of about 6-years-old."[26] The sling cradle was twined, as are many of the oldest textiles found in North and South America. Its inclusion with the grave goods left for a little boy suggests that it held significance for his culture. This parallels many early Andean burials.

Writing about the Navajo in 1938, W.W. Hill said, ". . . cornfields were defended at night by a slinger who threw stones about the field to scare them [coyotes, foxes and dogs] away."[27] The same practice in pre-Hispanic Peru was described and illustrated by Guaman Poma. (See figure 1.38.)

Of twenty-three North American tribes that used slings as toys, for scaring birds away from crops, and for herding sheep, there is evidence that nine to eleven also used them for battle.[28] (The term "toy" is somewhat misleading since it takes persistent practice to develop the skill of hitting a moving target with a slingstone.)

Evidence of the use of slings as combat weapons is found in the "stone balls" found at many Native North American archeological sites. Over the years, different ethnographers interviewed many Native Americans who related that while in recent times the sling was principally used for hunting birds and small game, for crop protection, and as a boy's toy, the "Ute, Navajo, Mescalero Apache, Yavapai, Hopi, Tewa (Walpi pueblo), Santa Ana, and San Ildefonso informants additionally claimed that the sling had functioned as a combat weapon."[29] Interestingly, slingstones with the same characteristics and age (up to 6000 years old) are as common north of the U.S./Mexican border as they are to the south, however, "in Mesoamerica such spheroids are often interpreted as slingstones, [but] in the Southwest they almost never are. Rather, they are usually and safely interpreted as 'function unknown,' with 'game balls' running a close second . . . ."[30]

# A Selection of Contemporary Slings from Around the World

Below are examples of working slings from various parts of the world. Most are less than 100 years old. They are made with vegetable and animal fibers and use a fascinating variety of construction techniques. (Andean slings appear in a separate section.)

## Saudi Arabia

In correspondence with Joy Totah Hilden, she made the following comments about two slings she photographed in Saudi Arabia, shown in figures 1.14 and 1.15:

> The little weaving on the sticks is intended as a sling-shot cradle. The warp is woven on the twigs; these are then removed, and the weft put in their place . . . [The slings] are woven from the fibre of the Salab tree. The stems are buried for 2 to 3 days, then the bark is removed and the fibres are shredded off. Total length of the sling rope is approximately one meter; the cradle is 5–6" long and 4" wide [14 cm × 10 cm].

FIGURE 1.14.
Cradle of a Saudi Arabian sling being woven on sticks.
*Courtesy of Joy Totah Hilden.*

## The Pitt Rivers Museum Collection

The Pitt Rivers Museum has a large collection of slings, a few of which are featured in figures 1.16–1.21.

FIGURE 1.15.
Saudi Arabian sling made of bast fibers.
*Courtesy of Joy Totah Hilden.*

FIGURE 1.16.
Spanish sling with a leather cradle and twisted rope sling cords.
*Pitt Rivers Museum, University of Oxford. 1927.67.10.*

FIGURE 1.17.
Algerian sling made with twisted strips of cloth in both the cords and cradle. *Pitt Rivers Museum, University of Oxford. 1913.17.38.*

19

**FIGURE 1.18.**
Southern Moroccan sling with top-stitching used to strengthen the cradle. *Pitt Rivers Museum, University of Oxford. 1956.12.26.*

**FIGURE 1.19.**
Syrian shepherd's sling with cords made from three groups of three threads each worked in 3-strand flat braids. The braid opens to provide nine threads for the cradle's warp.
*Pitt Rivers Museum, University of Oxford. 1933.90.35.*

**FIGURE 1.20.**
Mexican sling with 8-strand braided cords made of multiple strands of sisal. At the beginning of the cradle, the threads divide into smaller 3- and 4-strand braids that are opened to provide the warp for the pouch. *Pitt Rivers Museum, University of Oxford. 1917.53.258.*

**FIGURE 1.21.**
Sling from India made from soft rope. Multiple ends of twisted fibers form the cradle and the separately attached cords. *Pitt Rivers Museum, University of Oxford. 1884.29.15.*

## Slings in Pre-Hispanic Andean Cultures

Andean cultures were very sophisticated before the arrival of the Spanish; however, they had no system of writing that would help archaeologists and related specialists with dating slings and other artifacts. Also, depending on the time period when a description was written, and the specialty of the writer, the names or spellings of Andean locations, cultures, and time periods differ. We acknowledge that this causes confusion for the reader. Chronologies describing Andean cultures vary, in part, because new discoveries continue to be made and new technologies help refine our understanding of old ones. Direct quotes use the original terminology of the writer. Otherwise we have chosen to follow the style used by Jeffrey Quilter and some American archaeologists and ethnologists. For instance, we have used *Inca* instead of *Inka*, *Nazca* when describing a physical location on the Paracas Peninsula, and *Nasca* when describing the culture. (For a chronological table and map of pre-conquest cultures and sites see Figures 1.22 and 1.23.)

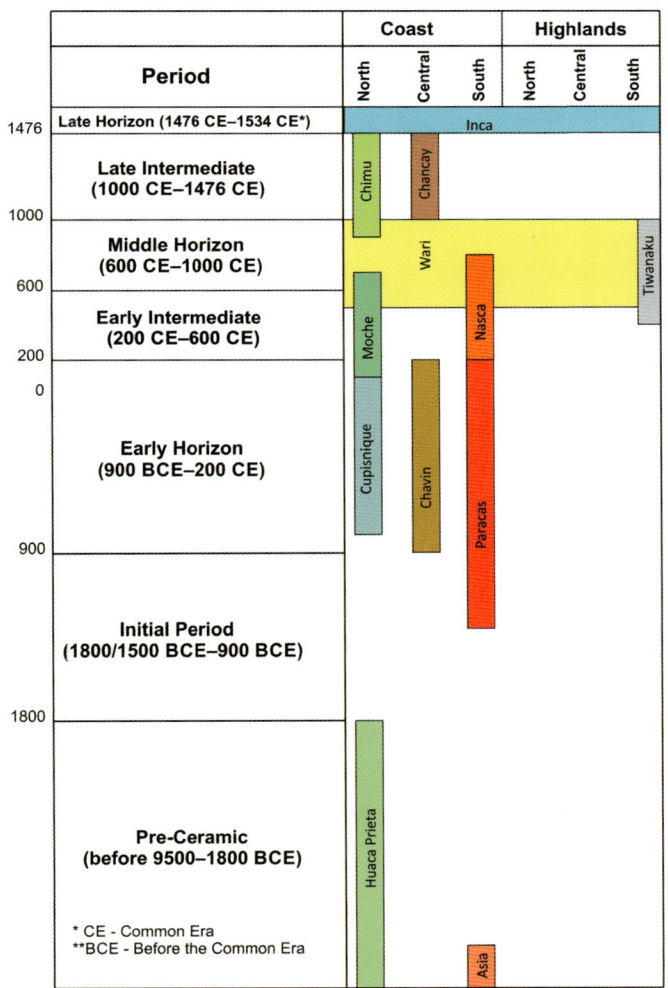

FIGURE 1.22.
Chronological table showing pre-conquest cultural periods in Peru.

FIGURE 1.23.
Map of Peru showing the location of pre-conquest sites.

Except for the stories collected by writers during the time of the Spanish Conquest of Peru, our knowledge and suppositions about pre-Hispanic Andean slings are built on the detective work of archaeologists, anthropologists, ethnologists, and specialists who have studied the artifacts of early Andean cultures. Remarkable treasures continue to be found throughout the Andes, buried on the extremely arid coast, in countless layers around Lima, and in the remote highlands. Slings have made their way into museums and private collections because for millennia the ancient Andeans buried their dead with honors, and slings were frequently part of grave goods.

This section describes a few of the cultures for which we have evidence of sling use, primarily coastal cultures whose burial goods were preserved in the arid coastal desert. Highland cultures are not as well represented because their climate is not conducive to preserving textiles. Instead, evidence for highland use is found in the defensive hilltop *pukaras* (fortresses) where architectural features, along with piles of slingstones, suggest defensive use of the sling. The belt of fortresses extends from at least northern highland Peru, throughout central and southern Peru, highland Bolivia, northern Chile, and into northwestern Argentina. Defensive siting, fortifications, and weapons such as slingstones and mace heads are common in many Late Intermediate and Middle Horizon sites.[31]

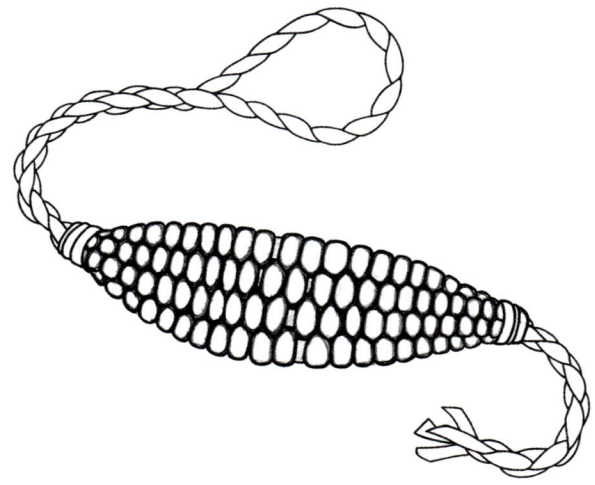

**FIGURE 1.25.**
*Early sling cradle from Asia, Peru, has no slit and features a loop to which a cord would be attached.*
*Drawing by Andy Owen.*

In the 1870s, when European and American interest in Andean archaeology first turned to Peru, archaeology was a new science. Methods for marking sites, describing layers, recording data, etc., were in their infancy. The Victorians were avid explorers and collectors, filling museums and private collections with objects from around the world. As the West became more interested in Peru's antiquities, the practice of grave robbing complicated the process of making sense of the "found" objects.

Grave robbing, known locally as *huaqueo*, from the Quechua term *huaca* (loosely meaning sacred place or object), dates back at least as far as the Spanish colonial period. During that time, mining concession grants were given for excavating precious metals from elite burials.[32] Because the Spanish were only concerned with exploiting precious metals, interest in the associated textiles and pottery remained local for about 200 years. Unfortunately, the practice of grave robbing continues for a variety of reasons. Looted Nasca graves often have large numbers of slings left behind because the huaqueros see little value in them.[33] Without information about a sling's exhumation, contextual clues that point to a sling's age or culture of origin are lost and slings easily become just textile objects. (Something their ancient owners would no doubt find strange.) Peru's National Culture Institute [now Ministry of Culture] continues to work to prevent damage to Peru's vast archaeological heritage despite overwhelming obstacles.[34]

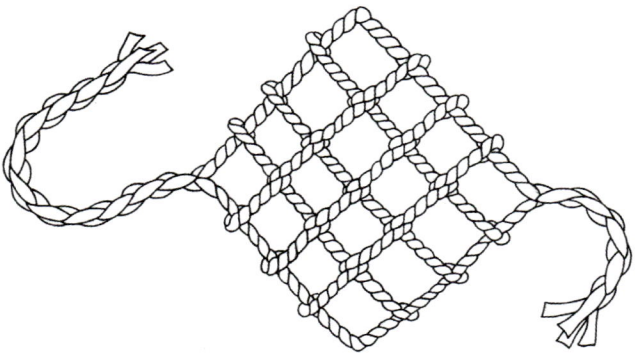

**FIGURE 1.24.**
*Early sling cradle, found at Huaca Prieta, made from plant fiber and plied cotton. The diamond-shaped mesh cradle is common in the slings of many later cultures.*
*Drawing by Andy Owen.*

*Huaca Prieta, Asia, and Otuma (3200–1200 BCE)*

In 1946, Junius Bird found early evidence of sling use at Huaca Prieta, a Late Pre-ceramic site (3000–1800 BCE) on the north coast of Peru. Life for these early inhabitants centered on the harvesting of fish and crustaceans from the sea and the labor of making the necessary fishing nets. "From all the material it is evident that throughout the long period of occupation, cotton was the primary fiber. An unidentified bast served as a supplementary or secondary fiber used in specific ways."[35]

Slings were made with *junco* (a sedge grass), *bast* fiber, or cotton with some slings containing yarns of more than one fiber.[36] *Bast* is a term that refers to fibers that can be extracted from the stems or leaves of a variety of plant species. In the Andes, bast fibers were obtained from the fleshy leaves of plants in the Agave family. Bast fiber slings are sometimes referred to as "maguey" although most likely the fiber came from local agave plants. The long length and strength of bast fibers made them particularly useful for slings.

Bird found cradles that were made with cowhitch knots sometimes combined with overhand knots. Figure 1.24 depicts a sling fragment from the Huaca Prieta excavation. It features a diamond-shaped mesh cradle made from junco attached to a plied bast cord.[37] (Only a few 3-strand braids were found.) This cradle design continued in popularity and is frequently seen in the slings of later cultures including the Chimu. Similar slings have been found in burials between the mummy wrappings and on the outermost layer, draped across the chests of men of high status, as seen in figure 1.26.[38]

At Huaca Prieta, few woven fabrics were found. "Throughout the occupation, finger-manipulated, twined cotton textiles predominated along with smaller, relatively stable amounts of weaving, netting, and looping."[39] Twining was used for constructing a wide variety of textiles including patterned fabrics. Many, including an iconic condor, are highlighted in Bird and Hyslop's book, *The Preceramic Excavations at the Huaca Prieta, Chicama Valley, Peru.*[40] This and other examples demonstrate that the Andean talent for creating complex cloth has a long history.

The drawing in figure 1.25 is of a sling cradle fragment found on the south coast of Peru at Asia, dated to 1200 BCE. It is made of sliced reed stems, compactly twined. The cradle, which measures 5¼" × 1½" (13.5 cm × 4 cm), has a loop at one end presumably for attaching the second cord. It does not have a slit. Identical slings were found at Rio Seco and at Otuma.[41] (See map in figure 1.23.)

FIGURE 1.26.
Open-mesh bast-fiber slings are often found draped across the chests of mummies wrapped in layers of complex and simple fabrics. *Copyright Bolton Council from the Bolton Museum Archive Collection.*

## The Paracas and Nasca Cultures

In 1925 archeologist Julio Tello discovered the source of elaborate textiles that were being offered for sale on the antiquities market. They were coming from the arid, sparsely populated Paracas peninsula.[42] The Paracas burials were excavated from three different zones: Cavernas, Arena Blanca, and Necropolis. They contained "the remains of a society that inhabited the region from approximately 600–175 B.C."[43] "The Paracas Necropolis cemetery, approximately 150 BC–AD 200 is the largest set of relatively well-preserved and well-documented burials of early complex societies on the desert coast of the Central Andes, one of the few regions of the world preserving evidence of textile history and its social contexts."[44] Several cultures used this burial ground and, "at nearly every site, two or more periods were represented among the remains."[45]

Like the earlier societies in Huaca Prieta, the residents of the Paracas peninsula depended on the sea. The Bahia de Paracas, a deep-water harbor, supported abundant marine life. Unlike at Huaca Prieta, weaving and the embellishment of textiles for garments and burial goods was of paramount importance to this culture. The desire to create textiles spurred agriculture; enormous amounts of cotton were necessary, as were bast fibers, for slings and other goods. Paracas craftspeople participated in long-range trade to obtain highly desirable goods: camelid fiber for embroidery and weaving; dyestuffs including cochineal for dyeing a wide range of colored yarns; spondylus shell for jewelry; tropical bird feathers for ritual paraphernalia.[46] "Even when measured against the high standards set by other pre-Hispanic Andean weavers, Paracas weavers excelled in their profession, both in the range of techniques exploited and in the mastery of their craft."[47]

Paracas and Nazca are very close geographically. Paracas is older, but there was a close relationship between the cultures. Slings were fairly common in the burials of both. They were placed in the layer closest to the body and sometimes in the outermost layer. The custom developed of renewing the cloth shrouds of high status individuals, adding additional layers of finely woven and embroidered cloth and ritual objects including slings.[48] In Anne Paul's analysis of Paracas Necropolis *Mummy Bundle 310*, seven of the ninety-two objects are bast fiber slings.[49]

Paracas slings were often made of bast fiber; some used cotton yarn, while others incorporated both. Most slings appear to be constructed by making the cradles and sling cords separately and then binding them together. Sling cords were usually started with a ¾" (2 cm) flat herringbone braid that was folded and bound to create the finger loop. Sometimes the finger loop incorporated dyed camelid yarn. (See figure 6.52 for instructions on making the herringbone pattern that is still used in slings and rope making.) The cradle was fashioned by making two lengths of herringbone braid, about 7" (18 cm) long, worked over thick cores. These were placed side-by-side and bound together to create a slitted cradle that was then attached to the sling cords.[50] (See figure 1.27.)

**FIGURE 1.28.**
Nasca core braid similar to one described by d'Harcourt. *Courtesy of the San Diego Museum of Man. Photograph by R. Owen.*

**FIGURE 1.27.**
Paracas sling cradle made of two herringbone braids worked over thick cores. © *The Field Museum, Catalog No CL0000_171289. Photograph by Sarah Rivers.*

**FIGURE 1.29.**
Poroma Nasca sling made with a technique called stem stitch by d'Harcourt. *Catalogue No 4.8360. Courtesy of the Phoebe A. Hearst Museum of Anthropology. Photograph by Ingrid Crickmore.*

**FIGURE 1.30.**
Late Pre-Inca. Dyed yarns are wrapped over the cord that binds the core. *Catalog No 4-9125.
Courtesy of the Phoebe A. Hearst Museum of Anthropology.
Photograph by Ingrid Crickmore.*

One example is described by Raoul d'Harcourt (his Plate 59 D) as "a pretty plaiting that was especially in vogue at Nazca. It required twenty-four strands of diverse thickness and of three colors and had a four-color variation."[53] The four-color Nasca braid fragment in figure 1.28 is a similar example that shows the skill of its spinner, dyer, and braid maker, likely the same individual. The red that outlines the diamonds is thinner than the dark green that zigzags down the length of the braid. The diamonds' eyes are filled with light green and yellow spun in two different thicknesses. (See *An Ancient and Contemporary Yarn Comparison in* Chapter 3 for additional information.)

The patterns in the sling cords shown in figures 1.29 and 1.30 were created by a technique called *wrapped weaving* by O'Neale[54] and *stem stitch* by d'Harcourt.[55] This is one of many instances in which the inventive constructions of early Andean textile artists are difficult to describe using Western systems for categorizing textile structures. The decomposed stitching on the sling cord in figure 1.31 reveals the base of the cord was constructed by carefully binding a bundle of core threads at frequent intervals. A thick cord was then wound in a continuous spiral around the core as a base onto which the pattern threads were worked. Figure 1.32 shows the diamond and zigzag patterns without a filling. (The cord may be in an unfinished state or the filling yarn has decomposed.)

**FIGURE 1.31.**
Pre-Hispanic sling cord from the Ica Nazca Region showing the structure that underlies the wrapped weaving.
*Catalogue No A30100.9.
Courtesy of the Smithsonian National Museum of Natural History.*

Slings that incorporated wool as a decorative element were found at Cahauchi, an early Nasca site. However, only a few were made exclusively of camelid fiber, suggesting that camelid fibers were still an expensive import from the highlands. Elaborate embroidery using camelid fiber yarns on plain-woven cloth appears to have been the focus of production rather than the complex textile structures that appear later in Nasca weaving and sling making.

Lila O'Neale, however, singles out one sling whose design was uncommon at that site, but is often seen in later Nasca slings. "Cahauchi 171314 was made of all-wool yarns probably llama wool, natural light cream and dark brown, dyed yellow and red." The cradle was woven in "Kelim tapestry technique too fine in quality for the rope-like ends."[51] The cradle was woven over 10 ribs of warp and featured a geometric pattern of yellow diamonds connected by yellow bars. The ends of the cradle are plain striped tapestry where it transitions to the slingcords that are formed from the cradle warps (ribs). The cords are made in "a technique called wrapped weaving, a term taken over from basketry . . . ."[52] The cords taper near the ends.

As presaged by the Cahauchi sling, during the Nasca Period (100 BCE–750 CE), slings become more decorative. Nasca sling makers used naturally-colored yarns combined with dyed yarns, or dyed yarns exclusively. The natural colors of the camelid fibers encompassed a wide range, pointing to the practice of selective breeding. Slings now incorporate complex braided, woven, or cross-knit looped structures. (See *Cross-Knit Looping* in Chapter 6.)

**FIGURE 1.32.**
Intact portion of the sling cord in figure 1.31.
*Catalogue No A30100.9
Courtesy of the Smithsonian National Museum of Natural History.*

During the Nasca Period slings became iconic, appearing as decorative elements on ceramic jugs and bowls and as decorative motifs. (Sling images are also seen on Moche ceramics.) The Nasca bowl in figure 1.33 features a sling worn as a headdress, a custom in many early societies. The focus of the bowl's design is the detailed painting of the sling and the colorful headbands beneath it. The cradle is similar to the Paracas cradle in figure 1.27. Donald Proulx, writing on Nasca ceramics, says:

∙∙∙∙∙∙∙∙∙∙∙∙∙∙∙∙∙∙∙∙∙∙∙∙∙∙∙∙∙∙∙∙∙∙∙∙∙∙∙∙∙∙∙∙∙∙∙∙∙∙∙∙∙∙∙∙∙∙∙

The importance of warfare in Nasca society is clearly seen in both the ceramic iconography and in the archaeological record. Nasca warriors holding clubs, or spears, and spear-throwers, and slingers are portrayed on some of the earliest Nasca pottery, often in association with human trophy heads. Slings are the most common weapon, being easily made and transported by even the most humble person.[56]

∙∙∙∙∙∙∙∙∙∙∙∙∙∙∙∙∙∙∙∙∙∙∙∙∙∙∙∙∙∙∙∙∙∙∙∙∙∙∙∙∙∙∙∙∙∙∙∙∙∙∙∙∙∙∙∙∙∙∙

FIGURE 1.33.
Finely painted Nasca bowl with a sling wrapped over a headdress. Ref. # PHM 4-8854. *Courtesy of the Phoebe A. Hearst Museum of Anthropology.*

*Ancón*

The Necropolis at Ancón, located about 25 miles (40 k) north of Lima on the central coast of Peru, was a burial site for many cultures, " . . . extending through all of the chronological periods recognized for the Central Andes."[57] There are two basic zones at the site, one dating from the Pre-ceramic to the Early Horizon (>1800–200 BCE), and another dating from the Early Intermediate Period to the Late Horizon (200 BCE–1550 CE).[58]

One of the early excavations made by Europeans was carried out in 1874–1875 by Wilhelm Reiss and Alphons Stübel, two German geologists who happened on excavations at Ancón and stayed to investigate. They described a variety of the slings and headbands in their richly illustrated book, *The Necropolis of Ancón in Perú: A Contribution to Our Knowledge of the Culture and Industries of the Empire of the Incas, Being the Result of Excavations Made on the Spot, Volume III*. They found maguey slings with diamond mesh-style cradles as well as wool slings with tapestry-woven cradles. They noted the difficulty of differentiating slings from "fillets" (headbands), and hypothesized about the frequent pattern changes found in the cords:

∙∙∙∙∙∙∙∙∙∙∙∙∙∙∙∙∙∙∙∙∙∙∙∙∙∙∙∙∙∙∙∙∙∙∙∙∙∙∙∙∙∙∙∙∙∙∙∙∙∙∙∙∙∙∙∙∙∙∙

It was seemingly the custom of the old Peruvians to use the slings as fillets, a practice still surviving there and in Bolivia, and which has the advantage of having the weapon always at hand in case of need. This twofold use of slings explains the occurrence of numerous transitions, so that it is often difficult to say whether the object is a fillet or a sling. They are mostly tightly and artistically plaited strings tapering towards both ends and with a special arrangement in the middle for the reception of the projectile. Two kinds of slings may be distinguished. In one the wide central portion itself is formed of the flexible strings used in making the whole weapon, and then a support is afforded to the stone by introducing a mesh . . . In the other, a thick stiff central piece of wool or even of leather is inserted . . . One end of the string is always looped for the finger while the other, which is freed with the fling, tapers to a fine point.[59]

∙∙∙∙∙∙∙∙∙∙∙∙∙∙∙∙∙∙∙∙∙∙∙∙∙∙∙∙∙∙∙∙∙∙∙∙∙∙∙∙∙∙∙∙∙∙∙∙∙∙∙∙∙∙∙∙∙∙∙

The question that Reiss and Stübel raised regarding the identification of sling versus headband is still valid. Some textiles identified as ceremonial slings may in fact have been headbands. Some may have been backstraps for looms.

**FIGURE 1.34.**
Pachacamac grass sling made with 3-strand braided cords and netted cradle of twisted and knotted fibers. *Pitt Rivers Museum, University of Oxford, 1884.29.13.*

*Pachacamac*

The sling in figure 1.34 was collected by Thomas Joseph Hutchinson at Pachacamac in 1873, twenty-three years before Max Uhle's historic excavations of the Pachacamac temple complex in 1896. There was no accompanying information to help determine its age; however, Pachacamac was occupied from the Middle through Late Horizons and perhaps earlier. The sling features 3-ply cords using multiple strands of grass. The cord fibers are divided for the netted cradle that includes a finely braided edging. A wooden handle is attached to one end.

*The Moche Empire*

The Moche came to power on the north coast of Peru in 100–700 CE. They are known for their exquisite ceramics. The north coast climate is too damp to preserve textiles and so evidence of slings comes mostly from the fine line paintings made on ceramic vessels. (See figure 1.35.) Moche fine line paintings frequently depict hand-to-hand combat weapons but also include distance weapons. The inclusion of "... slings, spear-throwers, and darts is testimony to a different type of combat.... In all warfare, ancient or modern, artillery or its equivalent is used to 'soften up' an enemy force before hand-to-hand fighting occurs."[60]

Slings and spear-throwers are ancient weapons that may have arrived with the first human inhabitants of the New World. Both weapons served not only military purposes but subsistence ones as well. "Slings are the traditional weapons of pastoralists to defend their herds and can also be used for hunting."[61]

**FIGURE 1.35.**
Slings are part of the story depicted in fine line paintings of Moche vessels. Drawing by Donna McClelland. The Christopher B. Donnan and Donna McClelland Moche Archive, Image Collections and Fieldwork Archives, Dunbarton Oaks, Trustees for Harvard University, Washington, D.C. *Courtesy of the Dunbarton Oaks Research Library and Collection.*

### *Tiwanaku and Wari*

Some clues to the existence of cultures that came before the Inca were discovered by the Conquistadors as they explored their new domain. The vast ruins of Tiahuanaco, located 9 miles (15 km) south of Lake Titicaca, Province of Ingavi, Department of La Paz, Bolivia, was visited by Cieza de León in 1549 but was otherwise unknown to outsiders until the mid-1800s. It was both a sacred site and the capital city of the Tiwanaku, whose culture flourished in the Middle Horizon (400–1000 CE). Evidence suggests they developed unique and sustainable agricultural practices in challenging environments that produced food for their people and for export. The Altiplano provided expansive grasslands for raising llamas for fiber, pack animals, and meat, and alpacas for their fine warm fiber. The fiber was also a valuable export item to coastal areas that were not conducive to raising camelids but where alpaca was highly valued. Slings would have been integral to herding and protecting llama caravans with their valuable cargos going to distant markets and returning with goods from lower altitudes. Tiwanaku influence spread from western Bolivia to southern Peru and to northern Argentina.

The Wari were a remarkable civilization that began in the Ayacucho region and spread its influence from the central Peruvian highlands to the coast between 600–1000 CE, during the Middle Horizon. Although the vast ruins of Huari, near modern day Ayacucho, had been visited by Cieza de León around 1550, it was not rediscovered until 1931 by Julio Tello who identified it as the Wari capital. "Ayacucho was the center of the Wari Empire, renowned for finely woven, sophisticated abstract tapestries . . . ."[62]

The Wari spread their influence over much of the central highlands and coast of what is present-day Peru. The southern edge of their empire bordered that of the Tiwanaku. Considered by some scholars to be the first empire in the Andes, there is no doubt that their influence was widely felt.[63] Given the expansionist nature of Wari and its military-related iconography and weaponry, some believe that Wari imperialism was concomitant with greater levels of violence relative to other pre-Hispanic groups in the Andes. While the soil conditions make Wari slings rare, weapons including slings were recovered from sites at Beringa and La Real.[64]

Many sites now designated as Wari were once thought to be Tiwanaku. As new discoveries have been made, some believe that the two cultures shared a religion and its iconography. Both cultures highly valued textiles along with ceramics. The sling shown in figure 1.36 was found on the south coast of Peru and may be Coastal Wari, 650–1000 CE. Its dimensions are 2½ × 1 × 99 inches (6.4 × 2.5 × 251.5 cm), obviously not intended for use in herding or warfare. Note the core-braided four-color diamond pattern on either side of the cradle.

FIGURE 1.36.
Ceremonial sling made of camelid fiber, found on the south coast of Peru.
*Courtesy Brooklyn Museum. Ref. 70.177.62.*

*The Inca*

The Inca were the last pre-Hispanic Andean culture to rule before the Spanish Conquest. They have been the most studied in part due to the accounts written during the conquest by the Spanish and by the first-generation indigenous people who translated for them. The origins of the Inca are uncertain and vary according to "who was telling the story, who was recording it, and when the tale was told, among other factors."[65] The Inca arose in the southern highlands in the thirteenth century. By the early 1400s, they were centered in Cuzco, coming to power rapidly and building an empire that "stretched from a small area of southern Colombia, through Peru, to include substantial parts of Bolivia, northwest Argentina, and northern Chile."[56] Their rule was very short, ending in 1576 when resistance to the Spanish was brutally crushed. Despite their relatively brief time in power, their civilization had many remarkable and unique characteristics.

*The Inca and sling myths.*

Many clues point to the importance of slings in Inca cultural rites and beliefs. Like the Nasca and Moche, the Inca referenced slings in the imagery used on textiles and ceramics. In addition to the visual record, there are a few written records. Most of these were made for Spanish administrators and clerics. One of the richest in terms of stories and drawings that include slings comes from Felipe Guaman Poma de Ayala, who was born into a noble Quechua family after the Spanish Conquest. His chronicle of life under Spanish rule, *Nueva corónica y buengobierno (The First New Chronicle and Good Government)*, finished in 1615, was intended to inform the Spanish king of the injustices and cruelty of the Spanish toward the Indians and to instruct the king in Inca social and political culture so that he could rule more wisely. (See figures 1.37 and 1.38.) Other major works from the period that also mention sling lore and military information were written by Pedro Cieza de Leon and Garcilaso de la Vega.

*The origin of the Inca.*

Inca creation myths incorporated sling use as did maturity rites:

About 18 miles (30 km) southeast of Cuzco, the founders of the Inca royal family, four brothers and four sisters, emerged from a cave. They set off on the road to Cuzco in search of new farmlands. During their travels, one brother, Ayar Kaci, climbed a hill where he cast slingstones at the hills with such force that he carved ravines where none had been before. He was greatly feared by his brothers and sisters so they decided to get rid of him. They lured him into a cave and sealed the entrance. Of the three remaining brothers, Manco, the founder of Cusco, was given instructions for performing the men's maturity rites.[67]

*Inca gods and rulers.*

In this land of extreme weather, the most important servants of the Creator in the Inca belief system were the sky gods, headed by the Sun who was believed to be the divine ancestor of the Inca dynasty.[68] While the Sun god had dedicated temples, Illapa, the thunder god was also very important: Illapa was understood to be:

. . . a man who lived in the sky and that he was made up of stars, with a war club in his left hand and a sling in his right hand. He was dressed in shining garments which gave off the flashes of lightning when he whirled his sling, and the crack of this sling made the thunder and he cracked his sling when he wanted it to rain.[69]

The rain was kept in a jug belonging to Thunder's sister, and he made it rain when he broke the jug with a well-aimed slingstone.[70]

The 8th Inca, Viracocha, was said to have once heated a slingstone until it was red hot and then hurled it across the Urubamba River in order to set fire to the thatched roofs of the town of Caytomarca.[71]

Legend says that with his right hand Pachacuti Inca Yupanqui (the 9th Inca) used his sling to throw a stone of gold at his enemy while defending himself with his shield in his left hand.[72] Poma's illustration, in figure 1.37, shows Pachacuti with sling and shield held as described for the god Thunder. Along with the mace, the sling and shield are symbolically associated with male maturity rites that were given to Manco, one of the original eight members of the Inca royal family.

**FIGURE 1.37.**
The 9th Inca, Pachacuti, with sling and shield held as described for the god Thunder. *Nueva corónica y buengobierno* (1615). Guaman Poma. *Courtesy of The Danish Library, Copenhagen.*

**FIGURE 1.38.**
Illustration of a farmer (sparrow guard). *Nueva corónica y buen gobierno* (1615). Guaman Poma. *Courtesy of The Danish Library, Copenhagen.*

## Historic Documentation on Military Use of the Sling

Although there is disagreement on the exact size of the Inca army, "All agree that armies numbering between 35,000 and 140,000 men were raised on various occasions with several large armies being in the field at once . . . Raising such large numbers of men without seriously disrupting the rest of Inca society and the economy required extreme care in planning."[73]

· · · · · · · · · · · · · · · · · · · · · · · · · · · · · · · · · · · · · · · · · · · ·

The Inca army was headed by the ruling Sapa (Emperor) Inca, who chose his generals from close family members. Ranking military personnel were chosen from specific clans. Lower ranking officers and troops drew on extended kinship groups, thus aligning the military with civic organization.[74]

. . . Troops were chosen from the common people who participated in the military as part of their labor tax obligation. "The choice of weapons for any particular unit was based on the traditions of the ethnic group of which it was comprised . . . There were whole units equipped with a single style of weapon; thus there were squadrons of archers, slingers, spearmen, and so on."[75]

· · · · · · · · · · · · · · · · · · · · · · · · · · · · · · · · · · · · · · · · · · · ·

The Inca army was highly disciplined according to accounts written by the Spanish, marching in ranks until given the order to attack. When attacking fortified positions, ranks of slingers and archers would begin the fight at long range. These troops gave way to ranks of hand-to-hand fighters using clubs and axes as well as short and long spear throwers.[76]

Military sling design and composition varied locally. Slings were commonly made from vegetable fibers, wool, or sometimes rawhide. The slings measured between 7" and 24" (18–60 cm) long when doubled. They featured a loop at one end and a cradle to hold a stone about the size of a hen's egg or larger.[77]

## Inca Maturity Rites for Boys

Guaman Poma's illustration in figure 1.37 shows the 9th Inca, Pachacuti, holding the sling, shield, and mace, three objects strongly associated with male maturity rites.

. . .at the end of each year in December boys of fourteen undertook the rite of maturity where they were given a breechcloth and a new name. The rites for ordinary people were probably simple ceremonies. For the royal families, however, the maturity rites for boys began in October when new clothes were made for their pilgrimage in November to Huanacauri. Each boy brought a llama to be sacrificed. The priests drew lines in blood in their faces and gave each boy a sling. A further stage of the rite was the beating of the boy's legs with slings to make him strong and brave. The rites ended when the boy's most important uncle gave him a shield, a sling, and a mace. At the end of the ceremony his ears were pierced for earplugs and he became a warrior."[78]

## Inca Agricultural and Hunting Practices

The illustration in figure 1.38 shows a "sparrow guard" protecting the crops with his sling. In Poma's chronicle, he advised the king on how farmers should guard their crops from planting time until the crops were ripe: ". . . guard the corn, the potatoes, the llamas well from the foxes and pumas. Guard the corn as well from the llamas, foxes, deer, skunks, . . . birds . . . and human thieves."[79] Hunting was strictly regulated to certain animals either for meat or for hair and fibers. Hunters used the sling, bola, nets for catching birds, snares, and nooses.[80]

# Andean Slings After the Fall of the Inca Empire

As the Spanish consolidated their power over the Inca and their empire, the Inca customs and regulations that required finely made textiles to dress the nobility, conduct religious ceremonies, and bury the dead, had to change. There was no longer a role for many of these complex fabrics. The practice of wrapping the head with fabulous braided or woven headbands topped with a handsome sling gave way to the wearing of hats. Much knowledge was lost as the skilled workshops disappeared; in the remote highlands, however, many textile traditions remained alive. Slings were among the textiles that still had a practical use. The sling braiders' art continued to find purpose in making complex braids for slings and dance/ceremonial costumes.

## Contemporary Andean Slings

Slings continue to be made and used in herding communities in the Andes. The following images depict contemporary slings (less than 100 years old) from the Owen collection. Additional examples are used throughout the book to illustrate various techniques. As with most slings that make their way to the West, information about their maker or the location in which they were made is usually unknown. Study them to appreciate the intricacy of the braids, and the skill with which pattern, color, and finishing choices were manipulated to create sophisticated designs. A chart, found in Appendix F, gives specific information on sling dimensions and construction. (See figure A.12.)

In figure 1.39, a Bolivian festival sling features a pattern of animals woven into the cradle that is edged with a decorative fringe. The medium blue in the fringe does not appear anywhere else in the textile and yet it adds just the touch to draw attention to the cradle.

The Ilave-style sling in figure 1.40 starts with a bicolored loop that was created by linking the white and brown threads and braiding each side of the loop independently before joining. The cords feature a striking diamond pattern that is punctuated by reversing spirals. It was made with handspun llama yarn in natural white, dark brown, and black. The cradle features a simple stepped diamond pattern. The second sling cord ends with four tiny 8-strand braids that are bound with a fine red yarn. A few red threads are added to each tassel.

Figure 1.41 is another Ilave-style sling, this one with zigzag and diamond patterns in the sling cords and a more complex tapestry cradle of stepped diamonds. Note that while the colors in the diamonds match, their backgrounds are different. In this cradle, the color play is subtle; in others, dramatic positive/negative color schemes are sometimes used.

The alpaca sling in figure 1.42 begins with a patterned loop, and continues with a simple design of white chevrons on a medium and dark striped cord. A short length of thicker diamond braid is added just before the cradle. The simple white diamond pattern on the dark background of the cradle echoes the initial patterns on the sling cords. Instructions for making an Ilave-style sling are found in Chapter 7, *Making an Andean-style Sling*.

FIGURE 1.41.
Andean sling with tapestry cradle and a variety of patterns on the sling cords. *Collection of R. Owen. Photograph by David Nut.*

FIGURE 1.42.
An alpaca sling with a simple diamond patterned cradle that begins with a handsome braided finger-loop. *Collection of R. Owen. Photograph by David Nutt.*

The cradle of the dance sling in figure 1.43 was woven with a yarn that was finer than that used in the sling cords to allow for weaving a more intricate pattern. (See figure 7.23 for a close-up of the cradle.) The cradle edging is chain stitch worked in multiple colors. The simple black and white patterned cords are finished with colorful tassels.

At first glance the sling in figure 1.44 appears to be constructed in the Ilave style, but instead it starts with the 24-strand sling cord. A blunt end was made by folding a warp of 12 strands in half, then using the combined warps in a 24-strand braid with an added 12-strand core. Then the warp was divided to create 12 warp groups for the cradle, which was woven with outlined diamond patterns. The warp was gathered for the second cord and an identical braid was made to match the first. Then the braid was rapidly decreased (some threads must have been removed) for a thin 4-strand braid. A warp was threaded through the cradle's beginning and the second 4-strand braid was made. Stitching covers the transition between the braid patterns. A narrow herringbone braid was stitched onto the cradle's edges. (See figures 6.52 and 7.22 for details.)

Another source of inspiring core braid designs comes from belt and headband braids. (See figures 5.1 and 6.40.) The braid in figure 5.1 may be a headband braid that Adele Cahlander described as being used by men in Huancavelica. It is worn wrapped twice around the head under a hat. It is finished at both ends with pairs of tasseled braids that hang down the back like hair braids. Both pre-Hispanic and contemporary braids are often embellished with tassels that can be simple or complex. (See *Inserted Tassels with Two Styles of Binding* in Chapter 6.) Additional colors, not found in the braid, are sometimes added to the tassels.

Additional information on the history of fibers and spinning is included in Chapter 3, *Core Braiding Yarns,* and further information about tassels and finishing techniques is found in Chapter 6, *Beginnings, Endings, and Embellishments.* See Appendix F, Comparison of Sling Dimensions, for measurements and additional information on the slings shown in the book.

**FIGURE 1.39.**
Fringed Bolivian festival sling with animals woven into the cradle. *Collection of R. Owen.*

**FIGURE 1.40.**
Andean sling with stepped diamond design in the cradle. *Collection of R. Owen.*

**FIGURE 1.43.**
Dance sling with finely woven cradle and tasseled sling cords. *Collection of R. Owen*

**FIGURE 1.44.**
Sling cradle with a pattern of outlined diamonds that are edged with a herringbone braid. *Collection of R. Owen*

FIGURE 2.1.
The core braiding frame set up for a 24-strand braid with a core of 12 strands.

# 2. BRAIDING EQUIPMENT

Core braids have traditionally been made in the hand. This book is written for two alternative methods: a core frame with a Japanese-style braiding stand and bobbins; and a hand-held braiding card. The core braiding frame offers several advantages over making core braids with the other methods. The braid is held under tension between the counterbalance weight, the active bobbins on the stand, and the core bobbins suspended on the harness. Thus, the bobbins do some of the work of tightening the braid structure as the braid is worked. This improves the smoothness of the braid at the exchange points where the core changes occur.

## The Core Braiding Frame

The core braiding frame is designed to hold a standard sized braiding stand beneath a harness that holds a set of core bobbins. (See figure 2.1.) The harness is made with two crosspieces that form the four arms. Each arm has three hooks and a slot for holding the core threads. The arms are attached to a dowel that can be moved vertically to adjust the working height of the core bobbins. The height of the braiding stand should be adjusted to work with the height of the braider's chair. As a general rule, the braider's forearms are roughly parallel to the floor when braiding.

For those who wish to make their own core frame, plans are found in Appendix A.

**FIGURE 2.2.**
All the braids in the book were made with 70 g bobbins.

## Bobbins

Core braiding uses Japanese-style braiding bobbins, also called *tama*. (See figure 2.2.) All of the samples in the book were made with 70 gram bobbins. Thirty-six bobbins will make all designs in this book but Design 32.1 that requires 40 bobbins.

**FIGURE 2.3.**
Bobbin with leader attached.

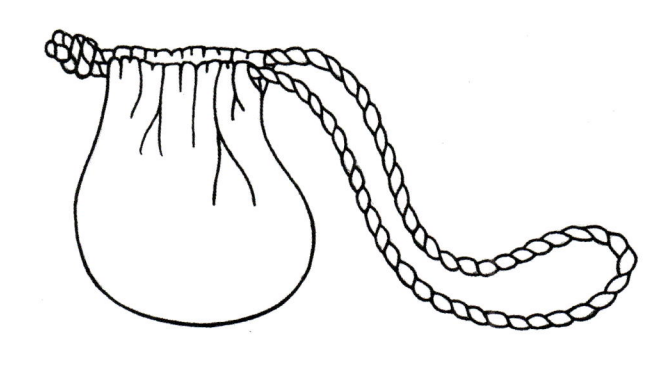

**FIGURE 2.4.**
Counterbalance bag.

## Counterbalance Bag

Make the counterbalance bag from a medium-weight woven fabric similar to quilting cotton. An average size for a finished bag is 5½" × 4½" (14 cm × 11.5 cm). (Avoid making the bag too long as it would have to be raised more frequently when braiding.) The bag has *two openings* in the casing for the drawstring. (See figures 2.4 and 2.5.) (If repurposing a bag with one slit in the casing, cut a second slit opposite the first.)

Make the drawstring from a tightly twisted cotton cord like rug warp or kitchen twine. The length of the drawstring should be no longer than 15–16" (38–41 cm). Make the bag as follows:

1. Cut a piece of cloth 12½" × 5½" (32 cm × 14 cm).

2. Make the casings on the 5½" sides by folding each raw edge over ¼" (6 mm) and pressing. Fold again ½" (13 mm) and topstitch close to the fold, as shown in figure 2.4.

3. With right sides together, fold in half and stitch side seams with a ½" (13 mm) seam. Stitch from the bottom fold toward the casing, *stopping 1" (2.5 cm) before* the edges of the casings. Backstitch to secure.

4. Thread a blunt end needle with the drawstring and pass it through the casings. Tie the ends together with a loose overhand knot. The bag will be attached to the braid with the *unknotted* section of the drawstring. Make a larkshead knot and check that you can insert the thumb and first two fingers into the loop when attaching the bag to a braid. Adjust the length if necessary. Tighten the knot. *Pull the knotted end close to the bag.*

## Warping Equipment

*Warp* is the name given to unbraided threads. The basic equipment used for measuring short warps is a pair of warping posts. (See figure 2.6.) These are clamped to a table. For longer warps, a third post can be added. A warping board (frame), used by weavers, can also be helpful if you frequently make long warps.

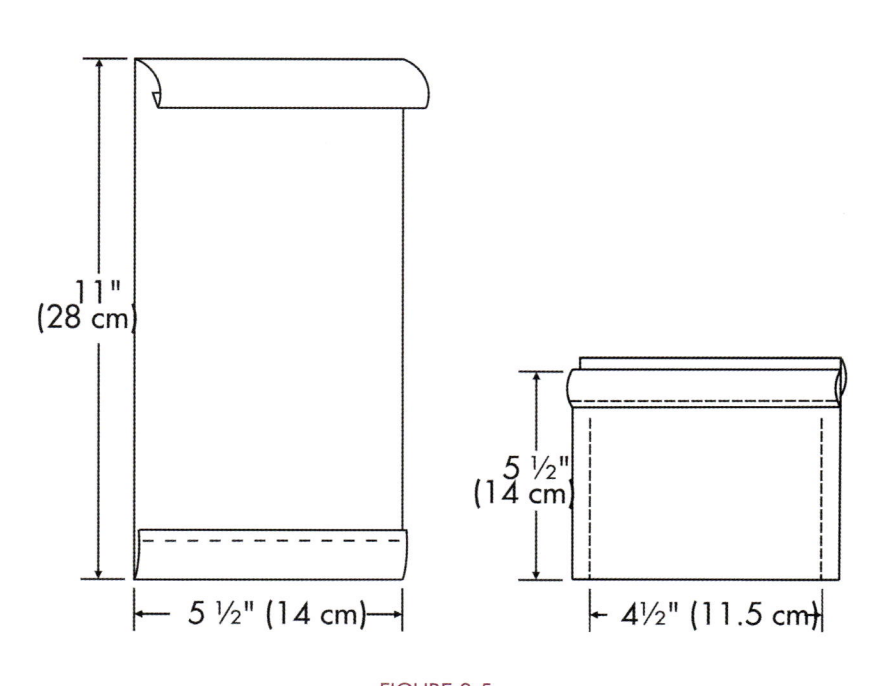

FIGURE 2.5.
The counterbalance bag is made from one piece of cloth with simple seams and casings.

FIGURE 2.6.
Warping post and clamp.

## Accessories

Freezer bag clips are very useful for preventing threads from slipping out of position as the braid is being set up. The two clips shown in figure 2.7 are 2" and 3" (5 cm and 7.5 cm) long.

The accessories shown in figure 2.8 are used as follows:

- A small S-hook is used when making braids that begin with a loop and also for initially holding the bag to the braid

- Sales tags are used for recording information about samples

- Crochet cotton (size 5 or 10) or 8/4 carpet warp is used for the "lifeline" threads described in Chapter 6, Chevron Braids, *Using a Lifeline*, and for temporarily binding the braid during the braiding process

- A wood or metal double-pointed knitting needle, approximate size US 6 (4.0 mm), is used when setting up the braid and when changing the height of the counterbalance bag

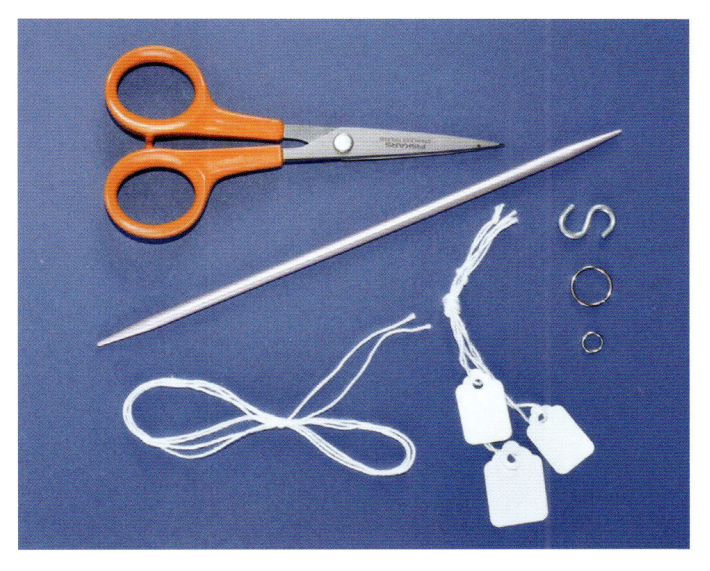

**FIGURE 2.8.**
Accessories that are helpful include: (*clockwise from top*) scissors, double-pointed knitting needle, S-hook, split rings, sales tags, and cotton binding thread.

Working the braid around a thin double-pointed knitting needle allows the creation of a more flexible braid with firmer stitches. This technique is useful when making trim for clothing. Sizes 1.5–2.25 mm work well. (See figure 2.9.) These needles are commonly used for lace and sock knitting.

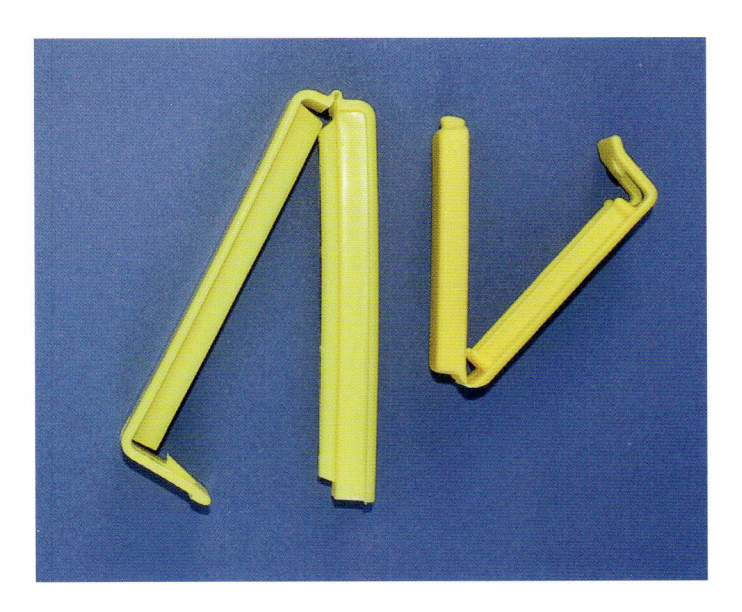

**FIGURE 2.7.**
Freezer bag clips are used to keep the threads from slipping when setting up the stand.

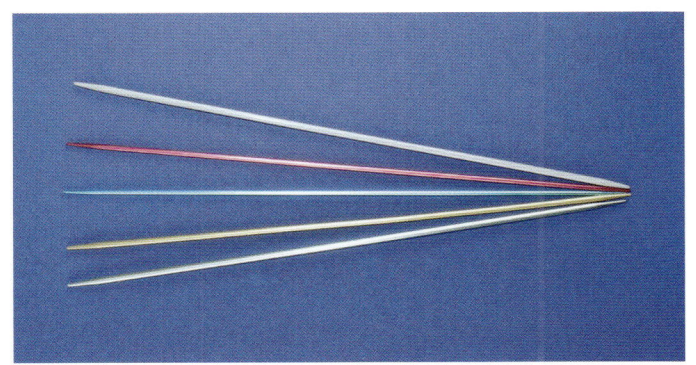

**FIGURE 2.9.**
Braids can be made over thin double-pointed knitting needles to increase the braid's flexibility.

## Counterbalance Weights

Compact weights such as sea fishing weights are recommended for use as a counterbalance as they suit the bag size suggested above. Large bolts, washers, or coins can be used as alternatives. You will need up to 1500 grams (53 oz.).

## The Braiding Card

The card is a versatile tool for making short lengths of braid or for testing designs without having to set up the stand. Long warps, like those for slings, can also be braided on the card. The measurements for the card shown in figure 2.10 are given in figure A.5 in Appendix B. (This card works for all of the designs with the exception of Design 32.1: those measurements are found in Appendix B in figure A.7.

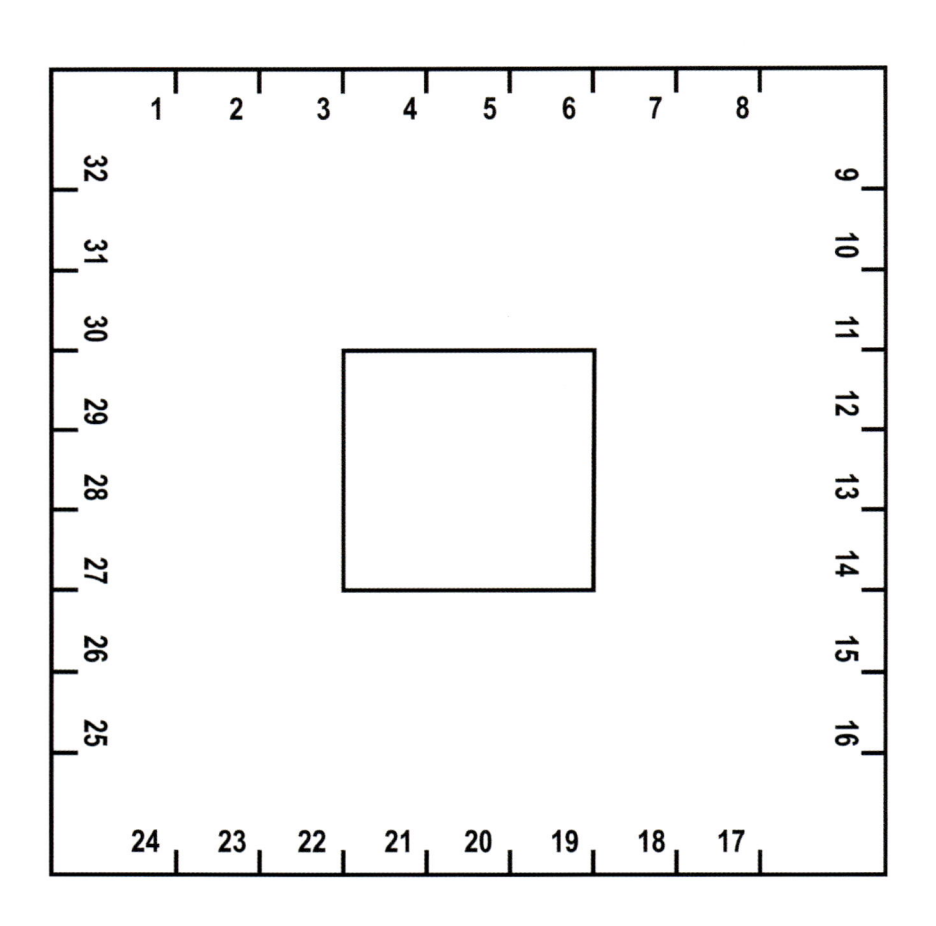

FIGURE 2.10.
Measurements for making a card designed for core braiding are given in Appendix B.

# 3. CORE BRAIDING YARNS:
## History, Contemporary Choices, and Design Considerations

The yarns used in pre-Hispanic sling braids were of course spun by hand. Contemporary herding slings are still made with custom-spun yarns while festival slings are made with handspun or manufactured yarns. The key feature of a good sling braiding yarn is the ability of the active braiding threads to cover the passive core threads. This chapter explores sling braiding yarns in both historical and contemporary contexts.

## Pre-Hispanic Fibers, Spinning, and Dyes

The earliest pre-Hispanic slings were simply constructed from cords made of twisted plant material. (See figures 1.24–1.26.) *These included grasses, cotton, and bast fibers from plants in the agave/maguey families.* Bast fibers, found inside the woody stems or stiff leaves of certain plants, were spun into both fine and coarse yarns used for a variety of textiles. For centuries, the primary sling design was made of plant fiber yarns (often maguey) and featured simply braided cords leading to a cradle. The cradles were either diamond-shaped mesh or lozenge-shaped braided or woven.

When highly decorative slings incorporating complex construction techniques like core braiding appeared in the Nasca period (100 BCE–800 CE), they were no longer made of maguey. They were instead made from firmly twisted, two-ply camelid fiber yarns. These yarns had qualities that allowed the construction of complex cord structures. Solid tapestry woven cradles replaced the mesh style. Over hundreds of years, the use of plant fibers for sling making eventually died out; there are few areas that use them today. The natural colors of the fleece predominated, but, as seen in the pretty little Nasca braid shown in figure 1.28, expertly dyed yarns were also used.

### Pre-Hispanic Camelid Breeds

Camelids originated in North America around 45 million years ago and are now found in South America, Asia, and Africa.[1] In South America, the guanaco is the wild ancestor of the llama. According to Junius Bird, the oldest wool yarns were spun from guanaco's superfine fiber. This animal ranged as far as:

**FIGURE 3.1.**
Wild vicunas grazing at Great Plains Reserva Nacional De Salinas. *Courtesy of Carole Thorpe.*

. . . the coastal zone of the region, and it would be predictable that cotton spinners would take advantage of such fiber whenever an animal was killed or captured. The same would be true when the ancestral wild forms of the llama and the alpaca were encountered. It is not improbable that desire for the wool was the major factor in llama and alpaca domestication . . .[2]

Vicuna, once an endangered species, are the wild ancestors of the alpaca. They are well adapted to the harsh climate, high elevation, and sparse vegetation of the Altiplano. (See figure 3.1.) More than half of their small population live in Peru.[3]

Domestication of most terrestrial mammals has resulted in hair color changes and a marked increase in the percentage of white hair . . . if one is to achieve clear, strong, and light shades of colors when dyeing, one needs white fibers . . . . The quantity and variety of dyed textiles that date to around 300–200 BC suggest that domestication of alpacas and llamas had succeeded in providing a good supply of white wool.[4]

**FIGURE 3.2.**
Llamas, like alpacas, come in a wide variety of natural colors.
*Courtesy of Carole Thorpe.*

The domestication of the alpaca and llama in the high elevations (3800+ meters above sea level) of the central Andean puna ecosystem of Peru began 6–7000 years ago. Camelid fiber yarns have been found in Andean weavings preserved at coastal sites dating to approximately 600 BCE–1200 CE.[5] Analysis of fibers from the ancient alpaca and llama mummies found at Yaral (700–1300 CE) support the belief that the practice of selectively breeding for color and fiber quality predates the Inca Empire. Several Yaral llamas had very fine, single coats containing no guard hairs; several had very coarse hairy coats; and there were a variety with other special characteristics.[6]

Under Inca rule, an annual census was taken of the state and shrine herds. Special emphasis was placed on breeding pure brown, black, and white animals for sacrifice to specific deities, as well as on quality fiber production for the state controlled textile industry and the production of sturdy pack llamas for the Inca army. Given such rigorous demands, it is likely that specific llama and alpaca breeds were maintained which subsequently disappeared during the Spanish conquest.[7]

Lila O'Neale, who analyzed the textiles from the Wari Kayan cemetery at Paracas (200 BCE–200 CE), is a good resource for information on the expertly spun and dyed yarns used in that early time. (See bibliography.) She described the weft in one early camelid sling. "The center is in Kelim tapestry technique, too fine in quality for the rope-like ends.... We can't be sure if a finer weft was spun for the cradle in order to allow the maker to weave a more intricate pattern, or if the finer weft improved the action of the sling."[8] The practice of spinning finer yarns for weaving more detailed patterns in the cradle is still seen in some contemporary Andean slings. (See figure 1.43.)

**FIGURE 3.3.**
White and colored alpacas, part of the breeding program at Mallikini Farm, Azangaro, Puno Province, Peru.
*Courtesy of Carole Thorpe.*

In commenting on the quality of the spinning in the Paracas yarns, O'Neale quotes D. C. Crawford:

. . . generalizations as to the quality of the old Peruvian yarns with few exceptions apply to the yarns in the . . . pieces of the Field Museum collections. Especially did he note the evenness of the spinning, the intentional insertion of a great degree of twist in some yarns, and the lesser amount of twist in others. The heavily twisted yarns give a crepe texture to the fabric, while the yarns less heavily twisted have "bite" enough to keep them in place in open voile-like fabrics.[9]

The "bite" is still an important characteristic of a good sling braiding yarn. Yarns for slings were, and continue to be, thicker than yarns customarily used for weaving. It is likely the yarns were spun using the stick spinning method used for slings today and not with the tiny spindles often found in workbaskets of the time.

## Pre-Hispanic Dyes

The weavers who created the elaborate textiles found in the Paracas and Nazca burials were skillful dyers; however, they preferred natural colors for some items.[10] Referring to the cotton slings found at Wari Kayan, O'Neale commented, "Animal fibers take dyes more satisfactorily than do vegetable fibers, but the fact that dyestuffs were available to afford a range of 190 colors suggests that Early Nazcans left their cotton undyed through choice rather than necessity."[11] Once camelid fibers were available, the preference for natural colors continued and encompassed a wide range from creamy white, to tans, cocoa browns, reddish browns, grays, and blacks.

Nasca braiders also incorporated dyed yarns into their slings and headband braids. Long before the Inca came to power, slings worn as headbands became an important item of dress. Some doubled as functional slings. Dyed yarns were popular in headband slings, particularly reds, often combined with natural color yarns.

Figure 3.4 shows a selection of natural dyes that were used in pre-Hispanic times. There has been a resurgence of interest in traditional dyeing and many Andean textile artists are seeking to expand the range of colors they can produce. Cochineal, in the first bowl on the left, is derived from a scale insect, and produces a rich range of colors from dark purple to reds and pinks. The bundle of stems on the left is indigo, a source of blues. Ch'illka leaves (last bowl on the right) produce mossy greens, and Q'olle and Kiko flowers produce a variety of yellows.[12] The skeins shown in figure 3.5 were dyed with those colorants. Some hues were created by over-dyeing with a second color.

**FIGURE 3.5.**
Skeins of handspun yarn dyed with a variety of natural dyes including cochineal and indigo.
*Courtesy of Carole Thorpe.*

**FIGURE 3.4.**
Natural dyes used in ancient and contemporary Andean textiles.
*Courtesy of Carole Thorpe.*

## Contemporary Andean Yarns for Slings

Andean herding slings are customarily made from handspun yarns made from llama fiber. The coarse longer guard hairs, which are customarily removed for garments, are often included in sling yarns. Natural colors predominate with dyed yarns occasionally used as accents.

• • • • • • • • • • • • • • • • • • • • • • • • • • • • • • • • • • • • • • • • • • • • • • •

Sling making is thought to be man's work, and men do their own spinning for it, frequently using a spinning method known in central Peru as *hilar a guayaquil* (guayaquil spinning) or *mismiy* in Quechua. This method is also used for rope making. A short stick, usually bamboo (the origin of the name guayaquil) in central Peru, is held in one hand, a roving in the other, and the two hands are rotated in opposite directions.[13] (See Nilda Callañaupa Alvarez's video on Andean spinning.[14])

• • • • • • • • • • • • • • • • • • • • • • • • • • • • • • • • • • • • • • • • • • • • • • •

A high degree of twist is put into the single strands. Two strands are then twisted in the opposite direction to create the finished 2-ply yarn. There are several methods of plying. (See Cahlander.[15]) The high twist yields a firm yarn that is strong and will resist abrasion when the sling is used. Another way to describe the texture of traditional sling yarns is that they have "tooth."

## Contemporary Andean Yarns for Festival Braids

Yarns for festival braids are thinner and are not coarse. They are made from alpaca, sheep's wool, or synthetic yarn. (Sheep were introduced by the Spanish and their wool is widely used in many Andean weaving communities where alpaca aren't raised.) The yarn may be handspun on a spindle or mechanically spun. It may be naturally colored or dyed with natural or synthetic dyes. A dance sling is an accessory to a costume; the emphasis is on the color and pattern, not on the fiber choice. Dance slings are often made of natural fiber yarns but in order to coordinate with the dance costume, manufactured yarns are sometimes used as well, especially in tassels and embellishments. (See figure 6.39.)

### An Ancient and Contemporary Yarn Comparison

For braiders who like to research ancient designs, Raoul d'Harcourt's book, *Textiles of Ancient Peru*, provides a unique source. The four color braid in figure 3.6 and the ancient Nasca braid shown in figure 1.28 have embossed patterns similar to the d'Harcourt's braid 63B.[16] In the author's (Owen) experience with reproducing braids from d'Harcourt's book, occasionally adjustments must be made to obtain his results. For instance, d'Harcourt described a core braid made of 38 strands of different thicknesses, with varying numbers of individual threads (elements) in each strand. After countless unsuccessful attempts to make this braid, the author reviewed the relationship of the numbers in the thread count. Table 5.1 shows that by halving the 16 yellow strands to make 8 strands, and adding 2 whites to make the 4 strands required for each eye, the 32-strand braid shown in figure 3.6 was made replicating d'Harcourt's braid. Table 3.1 compares d'Harcourt's thread count with ours.

Those interested in braid design will find that a variety of textured effects can be created by experimenting with the relative sizes of threads, just as ancient makers did. This can be done by grouping varying numbers of thin threads or by using different yarns.

**FIGURE 3.6.**
Thin threads were plied to create strands of varying thicknesses to replicate d'Harcourt's figure 63B.

**Table 3.1**

| d'Harcourt's Thread Count for 63B | | | |
|---|---|---|---|
| | color | elements | ends |
| 12 | yellow/brown | × 1 | 12 |
| 16 | yellow | × 3 | 48 |
| 2 | white | × 6 | 12 |
| 8 | red | × 4 | 32 |
| 38 | total strands | | 104 ends |

| Revised Thread Count for 24.13.1a | | | |
|---|---|---|---|
| | color | threads | |
| 12 | brown | × 1 | 12 |
| 8 | yellow | × 3 | 24 |
| 4 | white | × 6 | 24 |
| 8 | red | × 3 | 24 |
| 32 | total strands | | 84 ends |

## Yarn and Fiber Characteristics

Regardless of the braiding method, certain fibers and yarn styles are better suited to meeting the needs of a core braid's structure, i.e., threads must frequently change place from passive core to active braiding and back again. This requires frequent tightening and adjusting in order to help the stitches lie in the correct position and cover the core threads. This process stresses some yarns. The type of yarn chosen can make core braiding enjoyable or tedious. *Worsted-spun wool yarns that have a firm twist are the easiest to use for all three braiding methods (on the stand, on the card, or in the hand) and are highly recommended for beginners.* The majority of the samples shown in this book were made with tapestry wool. Our sample sling was made with a high quality wool/nylon rug yarn. There are a variety of other fibers and yarn styles that can also give pleasing results.

Yarn choice is influenced by the braiding method. Yarns for braids made in the hand, Andean style, need to have enough body and texture to hold the stitches in place once they are tightened. With experience, softer yarns like alpaca can be used. Braids made on a card are best worked using yarns that can withstand the abrasion caused by popping the strands in and out of the card slots. Braids made on the stand with weighted bobbins need plied yarns that have enough twist in the individual strands to prevent them from breaking when the weighted bobbins untwist the yarn. (This phenomenon is most noticeable when using alpaca and alpaca blend yarns that are undyed.) Despite this factor, the stand accommodates the largest variety of yarn styles. Before discussing the characteristics of yarns made from various fibers, it is helpful to understand a little bit about how yarn is constructed.

## Yarn Construction Basics

Spinning yarn requires three basic processes: preparing raw fiber, attenuating the fibers and adding twist, then winding the strand of yarn onto the spindle or bobbin. *Plying*—twisting two or more strands together—will usually be a subsequent process.

Whether spun by hand or mechanically, the way the fibers are prepared for spinning affects the characteristics of the finished yarn and its suitability for braiding. *Worsted* and *woolen* are Western terms used to describe distinctly different processes used to prepare the fibers before spinning adds the twist that transforms fibers into yarn. These terms are also commonly used to refer to finished goods made of wool but, for our purposes, we are using them as they pertain to yarn construction.

*Western handspun yarns.*

*Woolen-spun* yarns are very familiar to hand spinners. They are not suitable for braids in general or for core braids in particular. For woolen spinning on a spindle or wheel, the fibers are carded (brushed) to straighten them out and remove any bits of vegetation. This creates a batt in which the fibers lie in a roughly parallel arrangement. The batt is rolled up from the cut ends forming a *rolog*. As the fibers are twisted and pulled into the strand of yarn, they retain this spiral arrangement, giving the yarn superior ability to trap air and retain body heat. Two strands are then plied together in the opposite direction to stabilize the twist. However, the texture of the yarn is fuzzier than worsted-spun yarn (described next). More important, it will lack the strength necessary for the continuous process of tightening that core braids require.

*Worsted-spun* yarns, made from medium to long fibers, are superior for braiding whether hand spun or mechanically spun. One common way to prepare fibers for worsted handspinning in the West is to comb the fibers using wool combs and then pull them into a parallel arrangement called a *roving*. Twist is added, using a handspindle or a spinning wheel. Usually, two strands are plied together in the opposite direction to stabilize the twist. The parallel arrangement of the fibers, combined with a firm twist, creates a yarn that should withstand the friction of popping threads in and out of card slots or the stress of weighted bobbins and adjusting tension. Manufactured yarns are usually worsted spun regardless of the fiber.

*Andean handspun yarns.*

In traditional Andean spinning, a different worsted preparation method is used and applies to yarns made from wool, alpaca, or llama. The spinner makes a roving by first taking a handful of fiber and teasing out any clumps or snarls, making a cloud of fiber. These fibers are then pulled into a length of roving that is wound loosely around the spinner's arm as seen in figure 3.7. (For an excellent tutorial on Andean spinning, we recommend *Andean Spinning with Nilda Callañaupa Alvarez*.[17])

**FIGURE 3.7.**
Spinner making yarn in a small village south of Cusco, Peru.
*Photograph courtesy of Carole Thorpe.*

To begin spinning yarn, a few fibers are twisted in the fingers to form a strand of yarn that is attached to the end of the spindle with a half-hitch. The spindle is spun and the twist is allowed to run along a length of roving. The spindle is stopped, while the spinner feels along the new strand of yarn, pulling and thinning any lumps. When the diameter and uniformity of the yarn is correct, the spindle is given another twist to strengthen the strand. (Sometimes this is called *double drafting*.) The spindle is stopped, and the finished strand is wound onto the shaft and secured with a double half-hitch.[18]

The amount of twist put into the single strands depends on the purpose for the yarn. Knitting yarns have a looser, softer twist than weaving yarns but, in general, Andean yarns are more firmly spun than Western yarns for increased durability. All Andean yarns are plied to give them strength and are usually two-ply.

Plying takes two steps. First, the strands from two spindles are loosely plied onto a third (heavier) spindle by turning the spindle in the opposite direction of the original twist. The yarn is then skeined and washed and/or dyed. After drying, the yarn is spun again, putting in the necessary amount of twist for the particular project. Last, the yarn is wound into a very tight ball and left to rest so that the twist is set and the yarn will not un-ply when used.[19]

*Manufactured yarns.*

Manufactured yarns for knitting and weaving—regardless of the type of fiber—are usually worsted-spun and and plied with two or more strands. *To be suitable for core braiding they must also have a firm twist.* The exception is *Shetland*, *Shetland-style,* and some *homespun* yarns which are woolen-spun. If you want to pair a braid with a woolen fabric that you are going to knit or weave, choose worsted yarns in similar colors for the braid and make samples of both braid and cloth.

## Fibers for Braiding

Although llama, alpaca, and wool are the traditional fibers for braiding in the Andes, there are a variety of fibers that can be used including silk, cotton, rayon, linen, and blended yarns. In discussing the different characteristics of yarns, we describe wool first because it is the easiest fiber to use in making core braids, and it serves as a standard to which we can compare the characteristics of other fibers.

### Wool

Sheep and many other mammals have two-part coats composed of a soft downy undercoat that provides insulation and wicks moisture away from the skin, and coarser guard hairs that are water repellant and protect the undercoat. The majority of domestic sheep have had the hair bred out of their coat (except for the face and legs). Therefore, most require shearing. This is done yearly for fleeces that will be commercially spun; however, Andean spinners may let the fleece grow for two or three years to allow longer staple lengths for hand spinning stronger weaving yarns.[20]

The softness or scratchiness of a yarn is first dependent on the characteristics of the fleece, and second on the style of spinning. The fineness or coarseness of the fibers, the length of the staple (locks), and the amount of crimp, help determine what styles of yarn and purposes are appropriate.

The spinning term "count" refers to a measure of the relative fineness of the wool fibers. There are several systems for this including the Micron Count which is less subjective than older methods like the Bradford Count which is still in use and is useful for hand spinners.[21] *Merino* is an example of a breed with a fine count (<19–25 microns) They have a short staple length of about 2–4 inches (5–10 cm).[22] The wool has a high degree of crimp, which helps the yarn bounce back when stretched. Merino's softness and bounce make it a great choice for sweaters but a poor choice for braid making.

Most sheep in Peru are raised in small local flocks and are bred for a variety of characteristics that meet the needs of local weavers and knitters. (See figure 3.8.) Most are *Criollo*, descended from Spanish Churro and Merino sheep that have become well-adapted to the harsh Andean environment.[23]

Longwool breeds such as *Cotswold* and *Lincoln Longwool* possess fibers that have larger micron counts (35–38) and longer staple lengths—up to 12" (30 cm).[24] Their lustrous fleece was once prized in Britain's rug industry. If yarn from these breeds has a firm twist, it may make a good sling yarn. (See figure 3.9.) Breed specific yarns can be found at farms, or yarn shops that promote rare or specific breeds. Wool and fiber festivals, such as Maryland Sheep and Wool Festival, and regional sheep breeders' associations are good resources.

FIGURE 3.9.
A selection of lustrous hand-dyed English longwool yarns.

FIGURE 3.8.
Andean sheep are well adapted to their harsh environment.
*Courtesy of Carole Thorpe.*

The finer the fiber the more pleasant the wool will be next to the skin. Yarn manufacturers have become more particular about the qualities of the fleece they use, and it is not unusual to see a breed name on a label. The trend toward softer wools is great for knitters but not helpful for those looking for core braiding yarns. Another trend in hand knitting yarns is to treat wool fibers so that they are able to withstand the agitation and temperature changes of machine washing. These yarns, called *superwash*, are treated so that the scales on the wool fibers cannot lock together and shrink or felt. Superwash wools tend to be slippery, do not stand up to abrasion well, and are not a good choice for core braiding.

For sampling core braids, there are a wide variety of smooth, worsted-spun wool yarns available. Tapestry wool and Persian-style needlepoint wool come in a glorious range of colors making them easy to match to a project. (See figure 3.10.) They are made from long-staple fleece so they handle abrasion well, although they can be relatively expensive for large projects. Use the whole 3-ply strand of Perisan yarn for slings or purse handles or split it and use two strands for a braid about the diameter of one made with tapestry yarn.

**FIGURE 3.10.**
Tapestry wool and Persian needlepoint wool are excellent choices for core braiding and come in a wide variety of colors. *Photograph by Danielle Murphy.*

The following knitting wools are similar in thickness to tapestry yarn but are more economical to use and more pleasant against the skin than tapestry yarn: American *knitting worsted*, British *Double Knitting (DK),* and Australian *8-ply yarn*. Sold in 50–100 g balls, 50 g contains approximately 110–125 yards (101–118 m). The stitch gauge for knitters is in the range of 5–5.5 stitches per inch (20–22 stitches to 10 cm).

For sling making, wool or wool/nylon rug yarns are excellent choices.

The sample sling shown in figure 7.4a was made with *Collingwood Rug Yarn,* a high quality 80% wool/20% nylon yarn (900 yards/lb.). Be aware that weaving suppliers often offer rug yarns in mill end cones. To be suitable core braiding yarn they should all be the same size.

Handweaving suppliers may have a variety of worsted-spun wools that will be suitable for core braiding. These yarns will be less stretchy than many knitting yarns because weaving yarns are designed to have a higher twist and less stretch in order to handle the tension and abrasion of the weaving process. They are commonly sold in larger cones and skeins. Some suppliers offer smaller amounts.

## Alpaca and Llama Yarns

In general, alpacas produce higher quality fleeces than llamas. Alpacas typically have a single coat that contains almost no guard hairs. The llama, highly valued as a pack animal, usually has a higher percentage of guard hairs protecting its soft undercoat, which can also be valued for spinning. (There has been much hybridization in the Andes; some of today's alpacas have as much as forty percent hair in their fleeces.[25]) The fineness of the fiber also varies on different parts of a camelid's body. Although there are large growers, the majority of Peruvian alpacas and llamas are still raised on small family farms in the altiplano. (See figure 3.12.)

**FIGURE 3.11.**
Fine fleece huacaya alpaca at Mallikini Farm in Puno, Peru. *Courtesy of Carole Thorpe.*

There are two alpaca varieties. *Huacaya* (pronounced Wuh-kai-ya), sometimes described as teddy bear-like, has a crimpy wooly fleece and represents about 85% of the alpaca population in Peru. (See figures 3.3 and 3.11.) Huacaya has the widest range of natural colors. *Suri* alpaca, 15% of the Peruvian population and rare in other regions, has long, lustrous, distinctive locks that drape down the sides of its body.[26] (See figure 3.13.) Both varieties have exceptional thermal properties. Suri is more challenging to hand spin.

In Peru, yarn manufacturers classify fleeces according to the sight and touch of experienced fleece graders as well as by micron count. Interestingly, regardless of whether they came from llamas or alpacas, fibers in the 20-micron range are labeled as alpaca, and fibers in the 30-micron range are labeled as llama.[27] The alpaca fleece shown in figure 3.11 is very fine quality. Alpacas and llamas are increasingly raised in the USA, Great Britain, and in many other countries. Many growers offer fibers for spinners and some offer their own lines of yarn. Check to find out what grading system they use.

**FIGURE 3.13.**
Suri alpaca with their long silky fleece at Mallikini Farm in Puno, Peru. *Courtesy of Carole Thorpe.*

**FIGURE 3.12.**
Children help care for the family's animals on small family farms. *Photograph by Heddy Hollyfield.*

Even with very soft llama (or alpaca fleece), the yarn will be scratchy if guard hairs or coarser fibers are included in the yarn. (A yarn is only as soft as its coarsest fiber.) Less expensive yarns often include guard hairs that are undesirable in knitting yarns but are helpful in sling yarns. Alpaca and llama fiber have less crimp than wool and produce a less elastic yarn when mechanically spun. (However, Andean spinners find their handspun alpaca more elastic than their handspun wool.[28]) For braiders using bobbins, an advantage of mechanically spun alpaca and llama yarns is that they do not stretch under the weight of the bobbins, as do wool knitting yarns.

Mechanically-spun alpaca yarns usually have a medium twist and may be smooth or slightly fuzzy. If they are fuzzy, they will show braid patterns more clearly if there is enough contrast in the color value of the yarns (for instance, light gray with charcoal, instead of light gray with medium gray). Alpaca comes in a wide range of twenty-two natural colors from the characteristic pale creams and tans, to cocoa browns, reddish browns, grays, charcoal, and black, as well as in dyed colors. (See figure 3.14.) Buy enough yarn in one "dye lot" if you are making a garment since the natural colors are impossible to match.

Note that the shorter fiber length and medium twist found in mechanically-spun alpaca yarns sometimes cause difficulty when using single strands on the stand. In particular, some natural and heather colors are more subject to untwisting and breaking under the weight of the bobbins. (This can also happen with alpaca blend yarns.) Vat dyed colors don't tend to have this problem.

To prevent breakage, either use two strands of yarn per bobbin or, if this makes the braid too bulky, simply re-twist the strands of yarn after a given number of steps have been worked. This can be done by holding the thread with one hand near the edge of the braiding stand and, using the other hand, give the bobbin a twist. Let the twist run *up to the point of braiding*, lay the thread down, and re-twist the next strand. Repeat the process each time the counterbalance bag is moved up. This simple re-twisting routine is well worth the little bit of time it takes.

Large yarn companies do not offer many llama wool yarns but, occasionally baby llama or llama blend yarns are available. You may want to explore small producers.

**FIGURE 3.14.**
Mechanically-spun alpaca in natural and dyed colors is available in a variety of weights.

## Cotton

Cotton, a cellulose fiber, can be a good choice for core braiding although it does not have the body required for sling making. Cotton is indigenous to the Andean coastal region. In ancient times it had:

> . . . a staple which. . . ranges from an inch to an inch and three-fourths. The roughness which today makes it so valuable an ingredient for certain brands . . . of wool [blend] clothing, arises from innumerable tiny hooks which stand out along the fibres . . . . The colour of the cotton is white for the most part, but brown, tawny, and even blue cotton is frequently found in the work baskets contained in ancient burials along the coast.[29]

Mechanically-spun cottons have quite different characteristics from those described above by Phillip Means in 1927. The cottons best suited for braiding are *mercerized* and are designed for embroidery or weaving. Mercerization is a process in which the fibers are exposed to heat and a chemical with a strong base. This acts on the fibers to make them stronger and more lustrous. Gassing is an additional process whereby the yarn is passed very quickly over a small flame. This adds luster. These yarns come in a wide variety of styles and sizes, including: 6-strand embroidery floss, 10/2 and 20/2 pearl cotton for weaving, and #8, #12, and #16 pearl cottons for embroidery. (There are other sizes but these are the best suited for core braiding.) Most are available in a beautiful palette of colors. (See figure 3.15.)

### Other Fibers

For sling making, acrylic yarns and blends may be a practical choice. As in all fibers, acrylic comes in a variety of lengths and degrees of fineness. Look for yarns that are plied, have a smooth texture, and have medium to high twist. You won't know if they work until you try them. Feel free to experiment! When auditioning a new yarn, make sure you take the braid off-tension to check to see if you need to add or subtract counterbalance weight in order to get the stitches to cover the core and to check the flexibility of the braid.

**FIGURE 3.15.**
Mercerized cotton comes in a variety of sizes and a wide range of colors. *Photograph by Danielle Murphy.*

# 4. PREPARING THE WARP, SETTING UP FOR BRAIDING, AND BASIC WORKING METHODS

Chapter 4 covers the two most common methods for beginning a braid: starting with a tassel and starting with a larkshead knot. Chapter 6, *Beginnings, Endings, and Embellishments*, covers additional methods that are helpful in certain applications.

The warp is prepared in the same way for braiding on a stand or a card. The term *warp* refers to a group of threads that have been measured and prepared for braiding. (*Warp* can also be used to refer to one of those threads.) For braiding purposes, the term *strand* refers to a thread or a group of threads that are wound on one bobbin or are put in one slot of a braiding card. The term *end* is sometimes used to refer to a single thread as in "this 16-strand braid uses three ends of fine alpaca per bobbin."

When planning a warp, several things need to be taken into account: the desired diameter of the finished braid, the finished length, and additional amounts needed for tassels and special finishes.

## Planning the Braid Diameter

Depending on the yarn that is being used, there may be one end of yarn on each bobbin or in each card slot, or several ends. For 24-strand core braids, one extra end per strand can make a significant increase in the thickness of the braid, so sampling is highly recommended before starting a longer project.

## Planning the Warp Length

To calculate the warp length, take the desired length of the finished braid and add additional warp for tassels or special finishes and for *take-up*. Take-up is the additional warp used as the threads bend around each other, crossing the center of the braid as it is worked. For sampling:

- For 4-, 8-, and 16-strand braids add an additional 40–50% length to the warp

- For 24-strand core braids add an additional 50–60% length to the warp

When planning for a longer braid, always sample first. Use the sample to identify any short strands. This indicates strands that work more often and will need additional warp length so that they don't run out before the longer length is finished.

## Winding a Warp for Starting with a Tassel

Set the warping posts 18" (50 cm) apart for samples or to the distance needed for a particular project. (See figure 4.1.) Consider the type and length of the tassel you wish to have at the beginning of the braid. Add this length, plus an extra inch for trimming, to the warp length calculations. Two methods for winding warps are described:

Method 1. For warps with one strand per bobbin, one circuit around the warping posts yields one strand each for two bobbins.

Method 2. For warps with multiple threads per bobbin, one complete circuit (or more) yields the strands for *one bobbin*. The bobbin groups will be separated by twining. (See figure 4.2.)

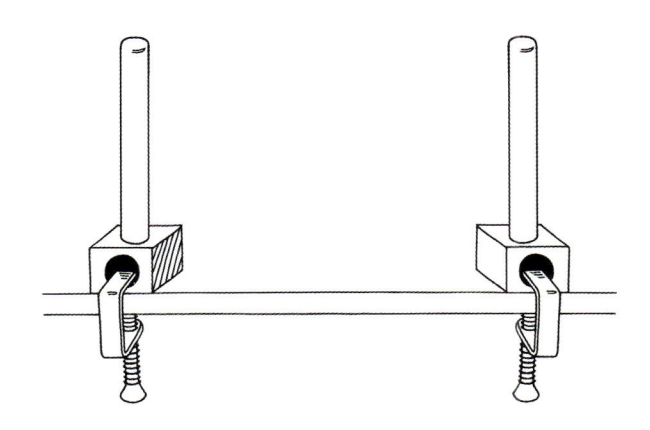

**FIGURE 4.1.**
Warping posts are clamped to a table.

The warp is wound in a continuous circular pattern beginning and ending at post A. (There is no cross in the warp as there would be for weaving.) The procedure is the same for both stand and card warps except for the tassel binding.

**FIGURE 4.2.**
Warp plan for a 16-strand 4-color braid. The solid vertical line indicates where the tassel binding will be made.
*Twining is only used for warps with multiple ends per bobbin.*

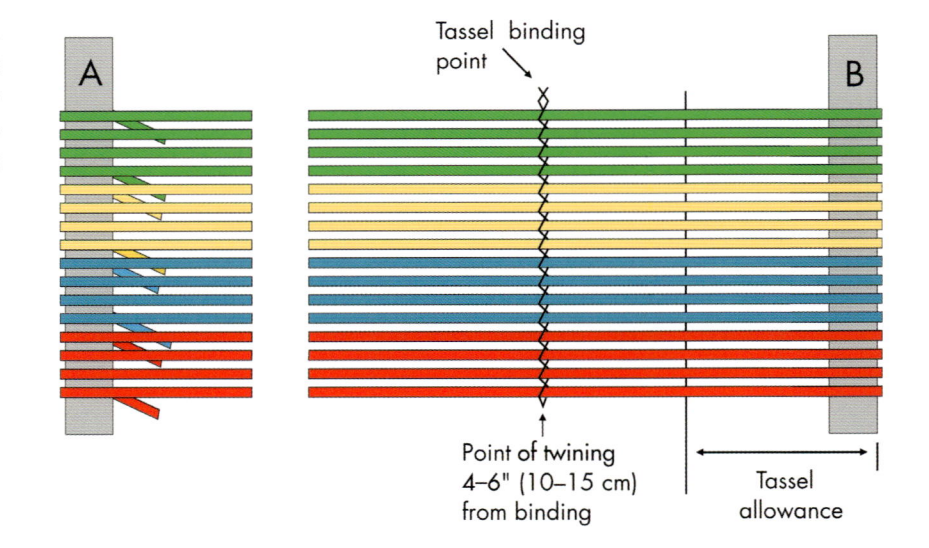

## Method 1: Warping for One End per Bobbin

1. Tie onto post A with color #1, wind around post B, and back to A. (See figure 4.2.) This creates warp for *two* bobbins (or two strands for the card).
2. Continue winding the pairs, counting the loops as you turn around post B, until you have wound all of the warps for color #1.
3. Tie color #1 off onto post A and tie on color #2 (or tie colors #1 and #2 together with an overhand knot placed close to the post). Wind the warp for the subsequent colors.
4. Secure the warp at post A. Twining will not be necessary. Skip to *Binding the Tassel*.

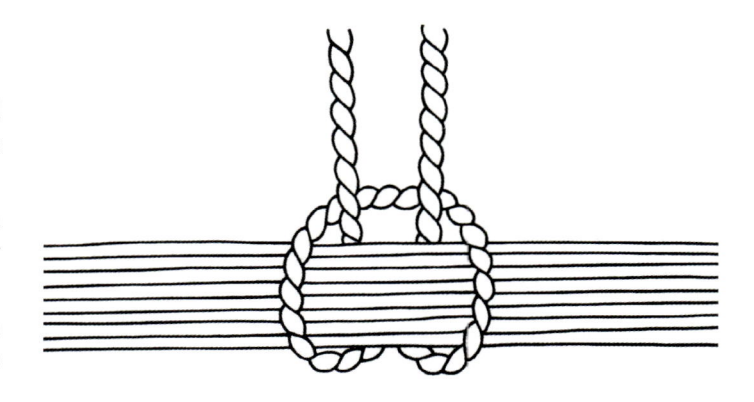

**FIGURE 4.3.**
Larkshead knot attaching the twining thread to the first warp group.

## Method 2: Warping for Multiple Ends per Bobbin

Twining keeps each bobbin group separate as it is wound and allows each group to be easily identified once the warp is removed from the pegs.

1. Tie color #1 onto post A. Wind around post B, and back to post A. This creates *two ends for bobbin #1*.

2. Continue winding pairs of threads, counting the loops as you turn around post B until you have wound all of the ends *for bobbin #1*. Wind the yarn several times around post A to keep tension on the warp while attaching the twining thread.

3. Cut a 24" (60 cm) length of twining thread. Fold it in half and make a larkshead knot (see figure 4.3) around the threads from both sides of the posts. Position the twining approximately 5–7" (13–17 cm) from post B.

4. Continue winding the next group of ends for bobbin #2. Secure the warp thread at post A and twine around the second bobbin's warp as shown in figure 4.4.

5. Continue winding each bobbin group, twining around each until all bobbin groups are complete. Tie the twining threads in a temporary knot (slipknot or bow knot).

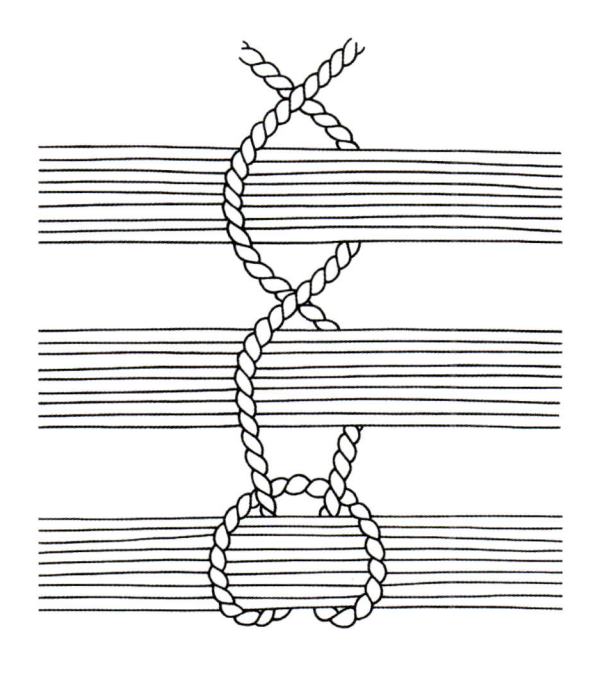

**FIGURE 4.4.**
Twining separates the bobbin groups, making them easier to count and separate when winding onto the bobbins.

**FIGURE 4.5.**
Measure tassel allowance from post B and bind in two places for insertion of the knitting needle when setting up the braiding stand, or one place if warping for a card.

## Binding the Tassel

Measure the tassel length back from post B, and bind the warp *snugly at one place for card warps and in two places for stand warps*. Figure 4.5 shows the warp bound and tied off in two places with a gap where the double-pointed knitting needle will be inserted to hold the warp on the stand while attaching the bobbins.

Tie off the bindings with double half hitches, square knots, or surgeon knots. Note: Pay attention to how the warp is bound. It must be tightly wrapped and tied while the warp is under tension on the posts. If tied loosely, the threads may pull out from the bindings, causing unevenness and slowing down the setting up process.

Cut the warp at post A. Remove the warp from the posts, and transfer it to the braiding stand or card as described in *Setting Up and Working on the Stand*, or *Setting Up and Working on the Card*.

## Winding Longer Warps

For making warps that are longer than a table's length, use three warping posts (see figure 4.6) or a warping board (frame) designed for weavers. Before attaching the posts to the table, use a tape measure to decide on the position of the third post. Clamp the posts in position and tie on to post A1. Wind to post A2 and then around post B. Continue back around A2, ending one complete circuit at post A1. Two warp ends will have been completed.

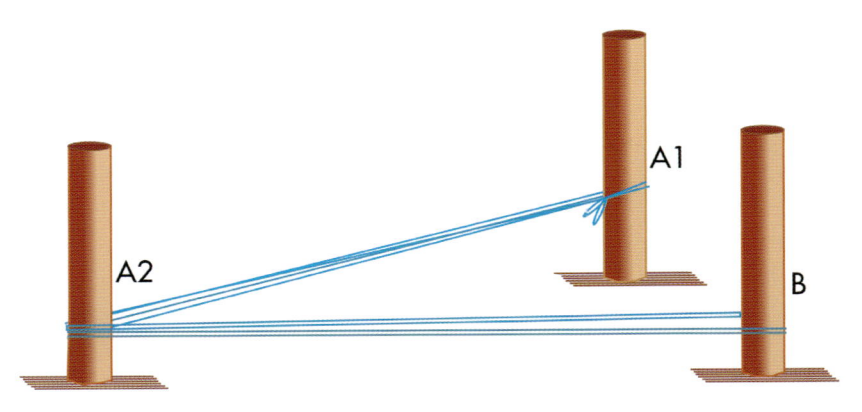

**FIGURE 4.6.**
Using a third warping post allows for making longer warps.

## Method A for Beginning with a Blunt End: Using a Larkshead Knot

Set up two posts and wind a circular warp. Make the larkshead tie by folding a 9" (15 cm) piece of binding string in half and tying the ends with an overhand knot. (See figure 4.7.) Insert the loop between the warp and post B. (See figure 4.5). Slip the knotted end through the loop and pull to secure.

Tie an overhand knot ½" (2 cm) from the larkshead knot to make a space for the knitting needle for stand braiding, or make the overhand knot further away from the larkshead to make a finger loop for card braiding. (See figure 4.7.)

**FIGURE 4.7.**
A larkshead knot secures the warps for beginning a braid with a blunt end.

## Setting Up and Working on the Card

Thread the tassel end of the warp down through the center hole of the card and hold the binding slightly below the top surface of the card. With the other hand, pull the threads into their respective slots according to the design's setting up diagram. (The threads will be centered on each side of the card with one or more empty slots on either side.) Adjust the tension on the threads so that the braid is centered in the hole.

*Tensioning on the card.*

When working on a card, the hands must do the work that the bobbins and counterbalance do when working on the stand. While moving the threads across the card to their new positions, the braid must be held under tension on the underside of the card. The tensioning is controlled by holding the braid firmly between two fingers while the threads are moved. One way to do this is shown in figure 5.15.

The order in which the threads are moved is shown in columns of numbers as seen in figure 5.5. This includes moves that will reposition the threads to their original setup slots before the next set of moves can be made.

## Setting Up and Working on the Stand

Thread the bound end of the warp through the center hole of the stand and insert a double-pointed knitting needle or chopstick between the tassel bindings. (See figure 4.8.) Pull the warp up so that the needle catches against the underside of the stand.

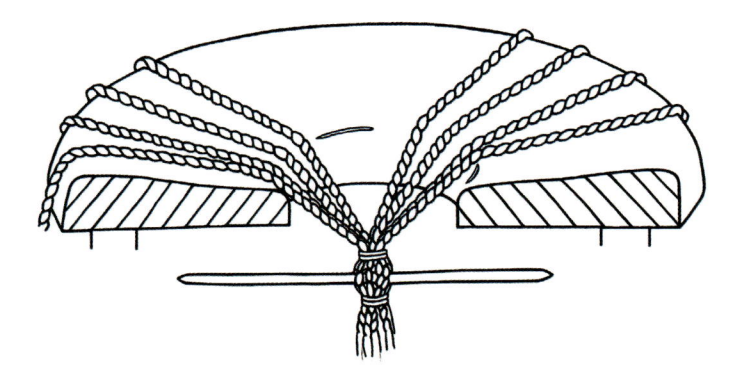

**FIGURE 4.8.**
A double-pointed knitting needle is inserted between the tassel bindings to hold the warp on the stand as the bobbins are attached.

*Attaching bobbins to the warp.*

A *weaver's knot* is used to attach the bobbins to the warp. This is a slipknot that is worked around the bobbin tie. (When braiding is finished, pull on the end of the tie to release the warp from the bobbin.) A *slipping hitch* is used to keep the extra warp on the bobbin while allowing the height of the bobbin to be adjusted.

*Making a weaver's knot.*

Select a warp thread and slip the tail through the bobbin tie loop. If right handed, hold the *tail of the warp* in the right hand and the *bobbin tie* in the left hand. Keeping the warp thread under tension, fold the tail back over itself, making a loop, and pull the loop through the opening as shown in figure 4.9.

Hold the loop in the left hand and pull it down tightly against the bobbin tie. The warp is now ready to wind onto the bobbin. (See figure 4.10.)

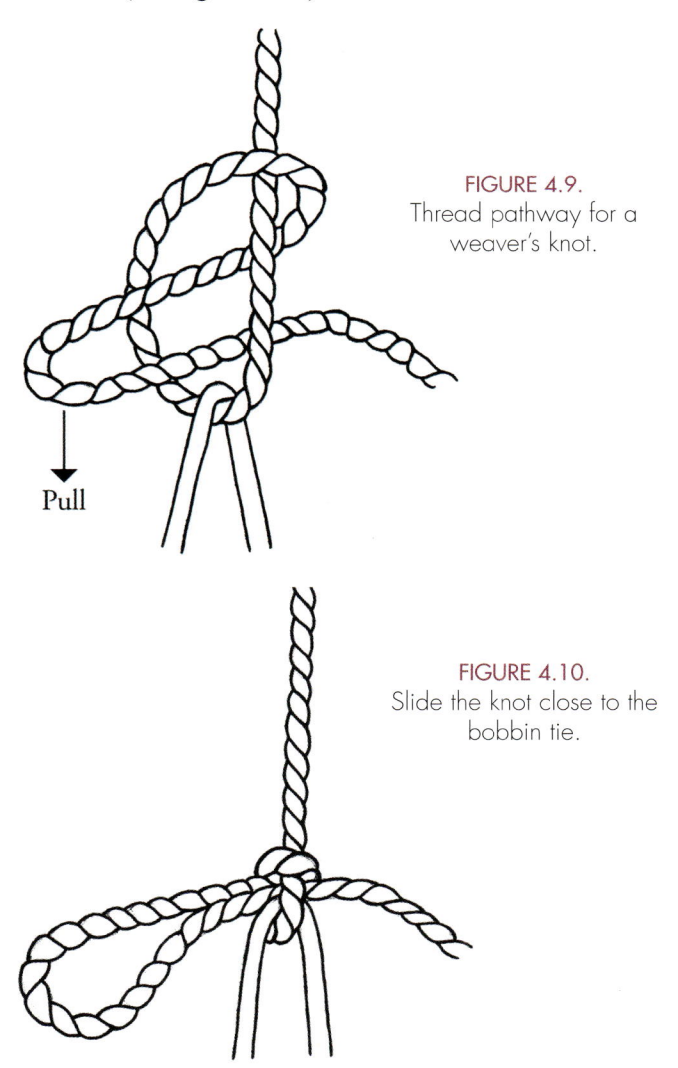

**FIGURE 4.9.**
Thread pathway for a weaver's knot.

**FIGURE 4.10.**
Slide the knot close to the bobbin tie.

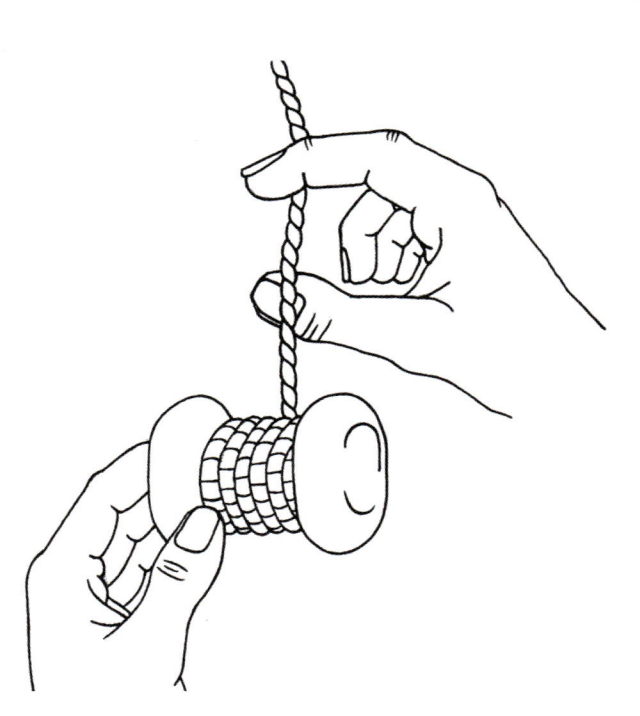

**FIGURE 4.11.**
Initial hand position for making slipping hitch with index finger in front and thumb behind the thread.

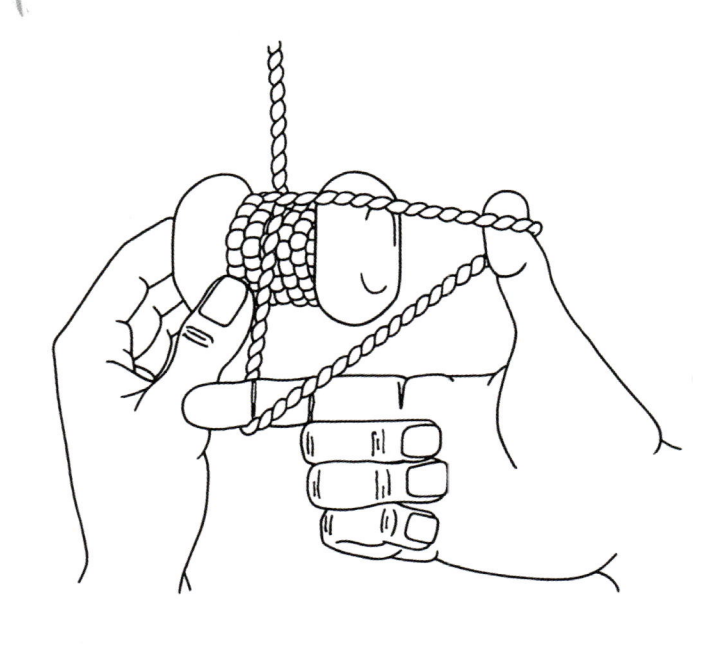

**FIGURE 4.13.**
Open the loop by turning the hand as shown and use thumb to lift the loop over the bobbin.

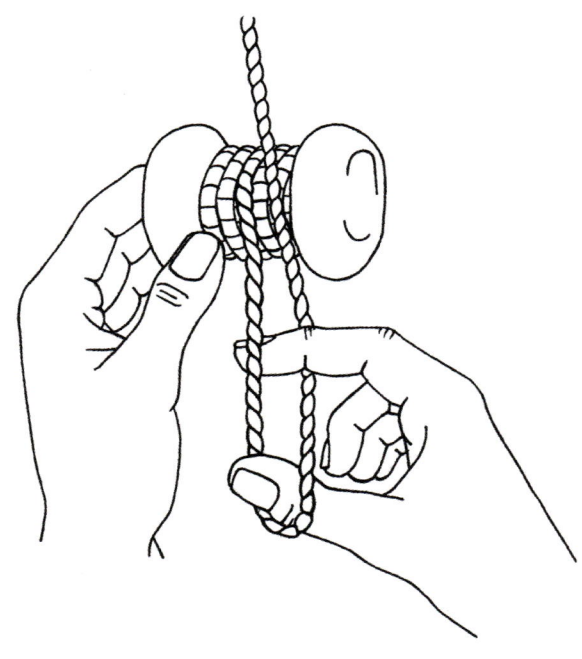

**FIGURE 4.12.**
Make the loop by swinging the bobbin under the right hand and away from the body.

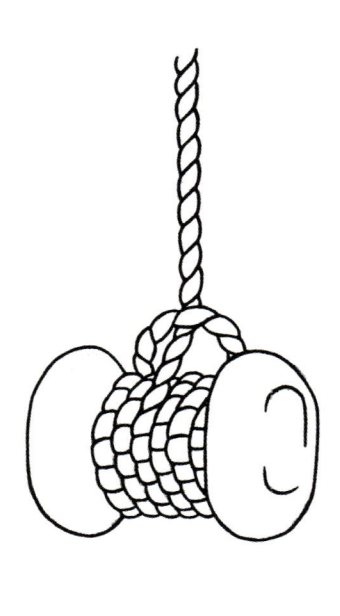

**FIGURE 4.14.**
Bobbin suspended by a slipping hitch.

## Setting Up a Three-color Core Braid Sample

A red and yellow braid with a blue core is used to demonstrate the sequence for setting up the core stand. The procedure for setting up braids with fewer strands is the same. (If you are new to working on the stand, practice making 8- and 16-strand braids before moving on to 24-strand braids. In this case, start with winding a warp of 8 threads for Design 8.1.1.) For the core braid sample, wind a warp with 12 strands each of red, yellow, and blue.

1. Separate the blue threads and lay them in the clamp.

2. Close the clamp and attach the bobbins. (See figure 4.15.) (If you are starting with a blunt end, as in figure 4.15, first match up pairs of threads and check that they are of equal length before clamping.)

3. Adjust the length of the bobbins so that they all hang about 12" below the top edge of the stand.

4. Open the clamp and place the core bobbins on the harness. (See figure 4.16.)

5. Select and clamp yellow strands. (See figure 4.16.)

6. Wind the first yellow bobbin. Bring down one core bobbin and adjust the length of the yellow strand to match it. Wind the remaining bobbins and adjust them to match. Return the core bobbin.

7. Remove yellow strands from the clamp and lay them on the stand. Clamp the red threads and attach the bobbins; adjust the length as before. (See figure 4.17.)

8. Arrange the red and yellow threads around the stand according to the set up pattern. (See figure 4.18.)

9. Attach the counterbalance bag and remove the knitting needle. (See figure 4.19.)

It is advisable to mark the end of each harness spoke with a cardinal point (N, S, E, and W). This will avoid confusion later when instructions call for rotating the core threads.

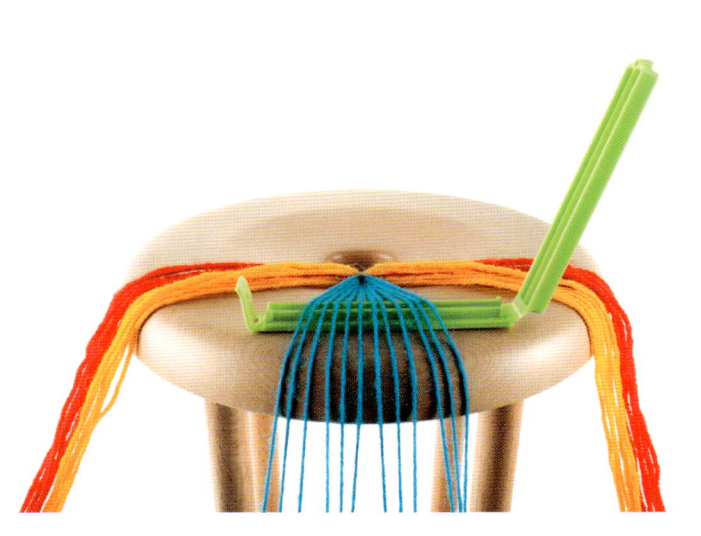

**FIGURE 4.15.**
Core bobbin strands are clamped, prior to attaching bobbins.

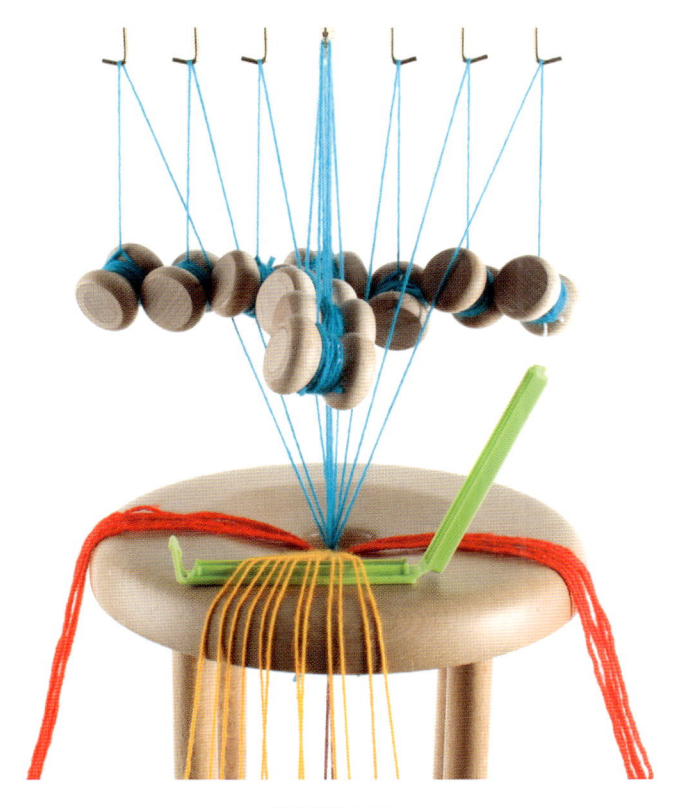

**FIGURE 4.16.**
Core bobbins are suspended before color A strands are prepared.

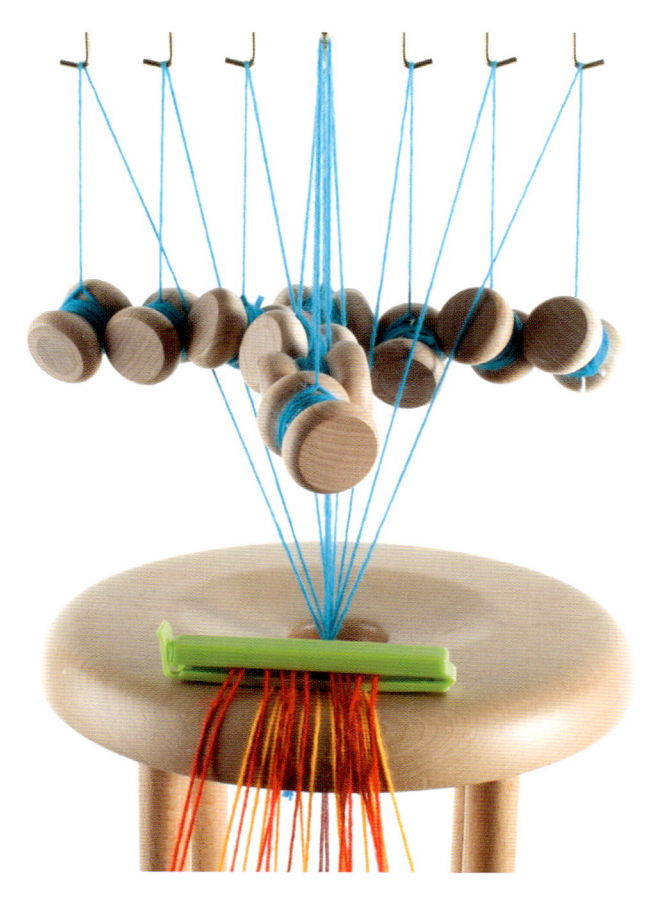

**FIGURE 4.17.**
Color B strands clamped and ready to have bobbins attached.

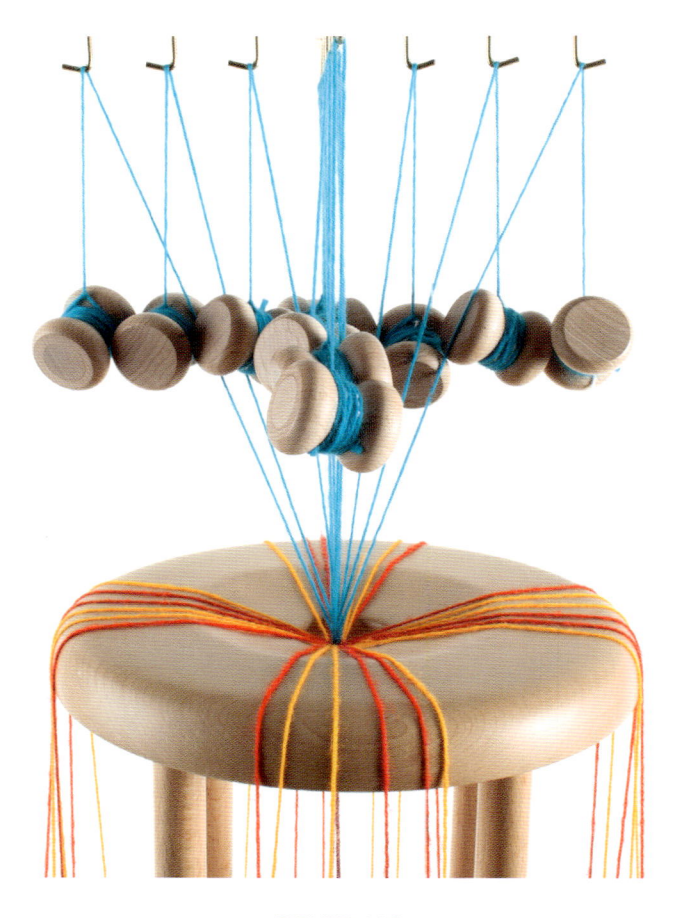

**FIGURE 4.18.**
The core stand with braiding stand completely "dressed."

56

## Choosing Counterbalance Weights

The choice of a counterbalance weight depends in part on the firmness or suppleness that is desired in the finished braid. Less weight will make a *firmer* braid; more weight will make the braid stitches longer and thus make the braid *more pliable*. The individual braider's method of tightening and the texture of the yarn may also influence how much weight is needed.

### Counterbalance Weights

As a guideline, calculate the weight of the counterbalance for a braid without a core at 50% of the total bobbin weight.

For braids with a core, as the braid is worked, a "push me/pull me" tension occurs between the bobbins pulling up on the core that is suspended on the harness and the counterbalance weight pulling the braid from below. The point of braiding should be located somewhere between the top and midway down the center hole of the braiding stand.

To adjust the tension on the threads, add or remove weights from the bag. There are additional variables that can affect the optimum weight: the number of bobbins in the core; the texture of the yarn; and the weight and the position of the bobbins hung on the harness. Determine the best counterbalance weight while making the sample to avoid tensioning irregularities in the braid. Average weights for 24-strand braids with cores of 12 strands are from 1100–1500 grams (39–53 oz.).

## Attaching the Counterbalance Bag

Place about half of the weights in the counterbalance bag. Use the unknotted end of the drawstring to pull the bag closed. Form a larkshead knot in the drawstring. Leaving your thumb, index, and middle fingers inside the larkshead, reach under the stand, pinch the area just under the knitting needle, and slip the larkshead knot between the tassel bindings. Hold it in place and tighten so that the bottom binding stops the knot from slipping down the warp. (See figure 4.19.) Open the side of the bag and insert the remaining weights. Remove the knitting needle and begin braiding.

After every 3–4" of braiding the counterbalance needs to be moved up. (Allowing the bag to "pendulum" makes braiding more difficult.) To prevent this, insert the knitting needle to stabilize the braid and adjust the height of the counterbalance bag.

**FIGURE 4.19.**
The counterbalance bag is attached to the warp with a larkshead knot placed between the bindings.

## Developing a Comfortable Working Position

The way in which the strands and their weighted bobbins are handled will influence the evenness of the braid. The braid designs in this book range from simple to complex. The braider's chosen method of working can make it easier or more difficult to make any given braid. One of the most important things you can do to improve your braiding is to be mindful of choosing a chair height that fits both you and the braiding stand. To avoid back strain, sit with your legs on either side of the stand and avoid chairs that have seats that slope back. For those who stand when braiding, experiment with different table heights to find the one that is most comfortable.

There are two basic processes and skills that must work together when a braid is being made:

1. The thought processes of selecting the correct strands to move, identifying their destination positions, and then visually checking to see that the new color sequence is correct.

2. The motor skills of lifting the strands, moving them across the stand while keeping them under tension, and placing them in their new position.

The strands are moved in pairs with the hands working in unison (with the exception of making core exchanges). As a strand is lifted, *the bobbin must remain suspended so that the strand remains under tension* until it is placed. Depending on the number of strands and the braid structure, there are two choices for how pairs of strands can be moved:

**Method 1.** A diagonal pair of threads (right-hand north thread paired with left-hand south thread) is moved as shown in Step 1 of figure 4.20. This style of movement is common in 4- and 8-strand braids and is used in some other braids. In order to efficiently move a diagonal pair of E/W threads, the hand position is "crossed" before the threads are moved. For example, in Step 2 of figure 4.20, the right hand reaches over the **top** of the stand to select the topmost strand on the west side of the stand while, with palm up, the left hand reaches across the **bottom** portion of the stand to select the lowest bobbin on the east side of the stand. The hands "uncross" as the strands are moved to their new positions.

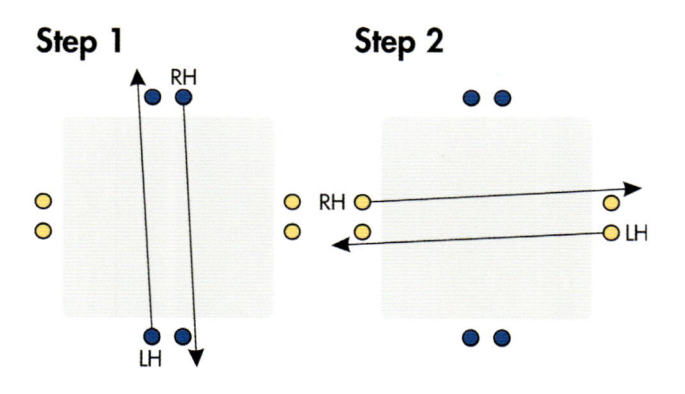

FIGURE 4.20.
Method 1. The active pair of threads are diagonally opposite each other.

**Method 2.** A pair of strands is moved from one side of the stand (south) to the opposite side (north); a new pair is selected and travels back to the original side (south) as seen in Step 1 of figure 4.21. The new pair is placed according to the design instructions. This style of movement is common in braids of 16 strands and more.

After the braider has identified which strands are to move and their destinations, the next choice has to do with how the strands are manipulated. This is determined by the strand's initial position.

- When moving threads **south to north**, put four fingers behind the strand of warp with the thumb lightly on top. Slide the hand down the strand until the little finger almost touches the bobbin. Lift the strand (with its bobbin hanging free), across the stand, rotating the wrist so that the palm faces down and the strand transfers to the thumb. With the free fingers, part the warps that are on the stand and lay the strand in position.

- When moving threads **north to south**, place the thumb under the strand of warp and slide the hand down until the little finger almost touches the bobbin. Lift the strand (with its bobbin hanging free) across the stand, rotating the wrist so that the strand lies across the fingers. With the free fingers, part the warps that are on the stand and lay the strand in position.

- The east and west threads are moved in the same manner.

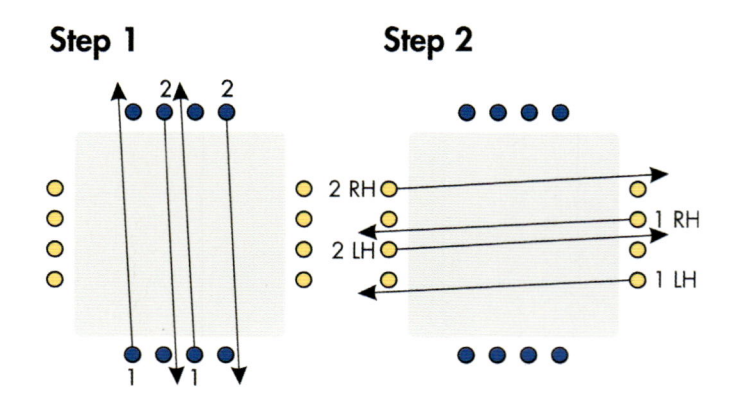

FIGURE 4.21.
Method 2. The active pairs of threads are moved from one side of the stand to the other.

# 5. BRAID DESIGNS 8.1–32.1

This chapter begins with two basic 8-strand braids, explaining how to interpret the diagrams used for the card and the stand. Then 16-strand braids that make spiral, striped, chevron, and diamond patterns are introduced. Last, 24-strand braids are explored, from simple chevrons to complex diamonds with multiple core changes. The braid seen in figure 5.1 hints at what you will discover—that as the number of strands increase, so do the roles of color and core changes in creating a wealth of patterns. (See Chapter 1, *Contemporary Andean Braids*, for more on this belt.)

## Eight-strand Braid Designs 8.1 and 8.2

Eight-strand designs, made from spiral and round structures, are commonly used in slings. Additional 8-strand designs can be found in previously published braiding books. (See bibliography.)

**8.1.1.**
Two-color clockwise spiral.

**8.1.2.**
Two-color anticlockwise spiral.

**8.1.3.**
Two-color reversed spirals.

### Interpreting Braiding Diagrams

Two styles of braiding diagrams are used for braids of 8 or more strands, *rotation diagrams* and *step diagrams*. (See figures 5.2 and 5.3.)

*Reading rotation diagrams.*

Rotation diagrams show the movement of the lower threads during a step. The four diagrams in figure 5.2 show the clockwise movements of Design 8.1.1 and the anticlockwise movements used in Design 8.1.2. Long arrows indicate the pathway of the *lower threads* (L) as they cross to the opposite side, to lie on top of the braid structure, thus changing their designation to *upper threads* (U). Each L thread that crosses completes a stitch in the braid on one side and begins a stitch on the opposite side. Short arrows point to the new position of the threads that shift to the right or left to make room for the new upper threads. With each move, the shifted U threads become L threads in the next braiding step.

**FIGURE 5.1.**
This headband or belt braid displays myriad diamond and chevron variations and is finished with braided tassels.
*Collection of R. Owen.*

*Reading diagrams for the stand.*

On each Design page, *step diagrams* show the initial color order of the threads with each strand of yarn represented by a colored circle. The beginning and ending destinations of the threads that will be moved are shown by arrows. (Later, in braids of 16-strands and more, it will be important to remember that after a strand has moved, the arrowhead represents the strand's new position.) (See figure 5.3.)

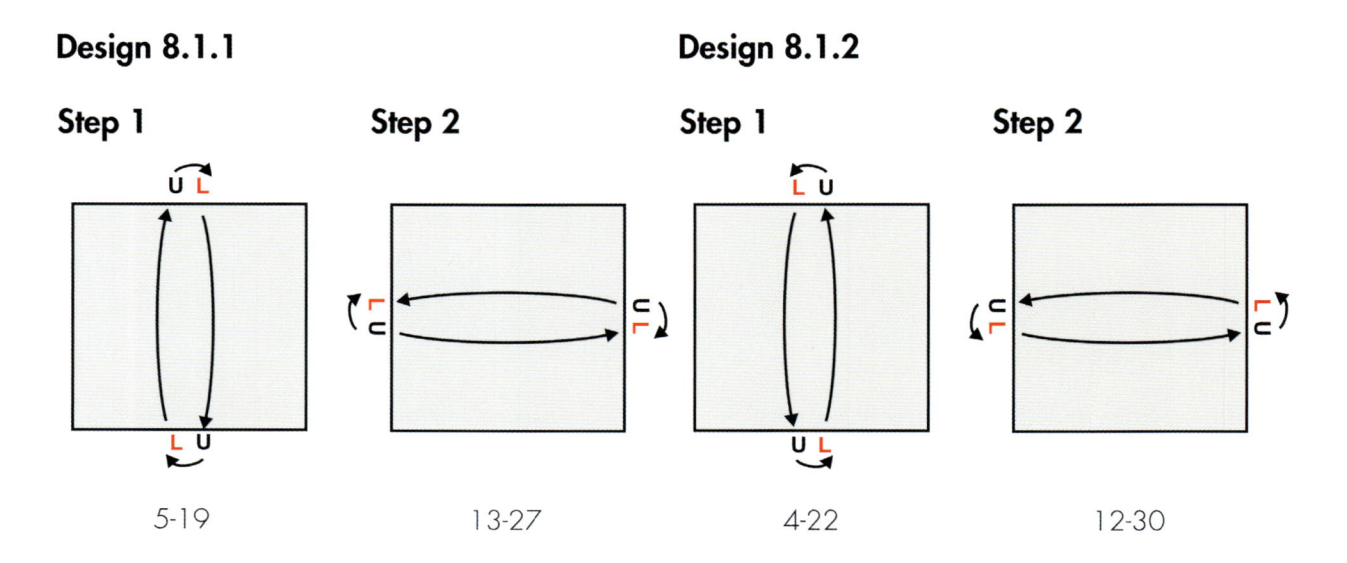

**Design 8.1.1**

Step 1 — 5-19

Step 2 — 13-27

**Design 8.1.2**

Step 1 — 4-22

Step 2 — 12-30

FIGURE 5.2.
Rotation diagrams show the clockwise and anticlockwise movements of the North/South (N/S) and East/West (E/W) threads. The diagrams above correspond to Steps 1 and 2 of Design 8.1.1 and Steps 3 and 4 of Design 8.1.2, seen in figure 5.3.

**Design 8.1.1**

Step 1 — 5-19

Step 2 — 13-27

**Design 8.1.2**

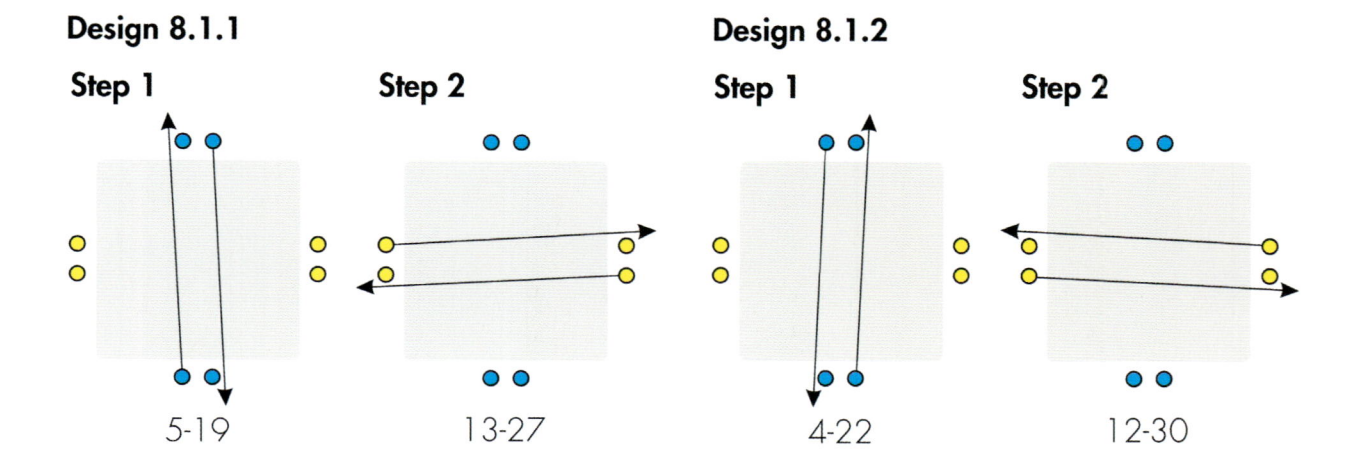

Step 1 — 4-22

Step 2 — 12-30

FIGURE 5.3.
Step diagrams use circles to show the color arrangement of the strands and arrows to indicate their destinations.

*Reading diagrams for the card.*

*Home slots,* shown in red in figure 5.4, are the center slots in which threads are placed when a card is set up for braiding. These are also the slots to which all threads return after completing the moves for each step. There must be two additional *working slots* on either side of the home slots to hold a thread while another thread moves out of its path.

The instructions for making a set of braiding moves on a card are written in two ways:

**Method 1.** Numbered columns, as seen in figure 5.5, will appear in some designs in addition to the stand braiding diagrams. To read them, start at the top of the left column, follow the N/S moves, and then use the right column to make the E/W moves. When finished, reposition the threads in the home slots to ready them for the next step.

**Method 2.** Key moves: Because rotation moves are used so frequently in braid patterns, they are quickly memorized. The term *key moves* refer to the top pairs of card move numbers in each column of rotation moves. The key moves, shown in red in figure 5.5, denote a sequence of clockwise or anticlockwise moves. In some designs, the complete card moves will be shown in two columns, as in figure 5.5. All designs, however, will show the key moves beneath the stand diagrams as in figure 5.6.

**Clockwise**

| 5 -19 | 12-30 |
| 4 - 5 | 13-12 |
| 21- 4 | 28-13 |

FIGURE 5.5.
*Rotation moves* for Design 8.1.
The key moves are shown in red.

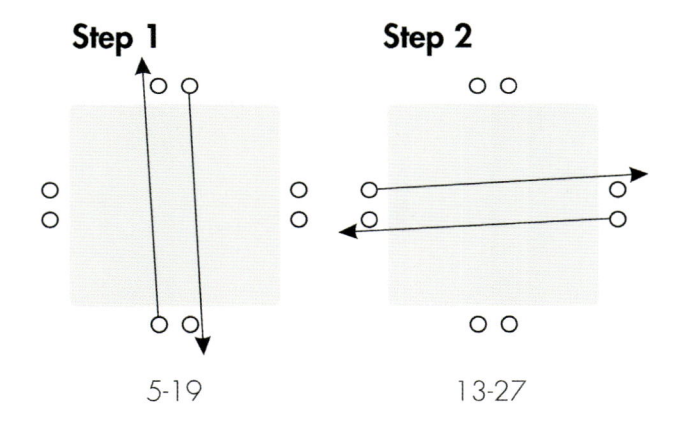

Step 1    5-19

Step 2    13-27

FIGURE 5.6.
The *key moves* for the card are shown under each step diagram for the stand. The arrows in the diagram indicate the threads that are to be moved and where they are placed.

Finishing with a Tassel

To keep the braid from unraveling, thread a tapestry needle with a warp thread. Wrap the thread firmly around the braid two or three times. Stitch through the braid, just above the wrapping, coming out at a stitch of the same color. Give a little tug. Then stitch diagonally through the braid, bringing the needle out in the center of the tassel. Trim the tassel to the desired length.

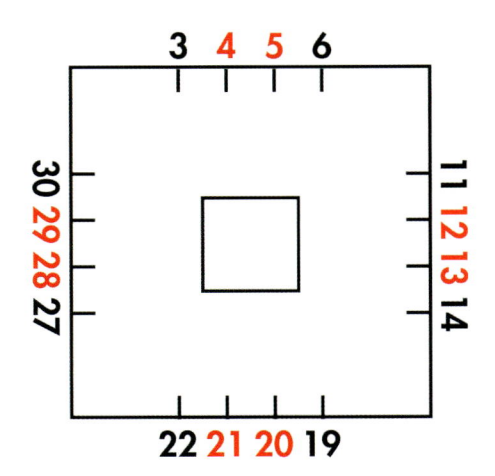

FIGURE 5.4.
Card *home slots* (red) and *working slots* (black) for Designs 8.1 and 8.2.

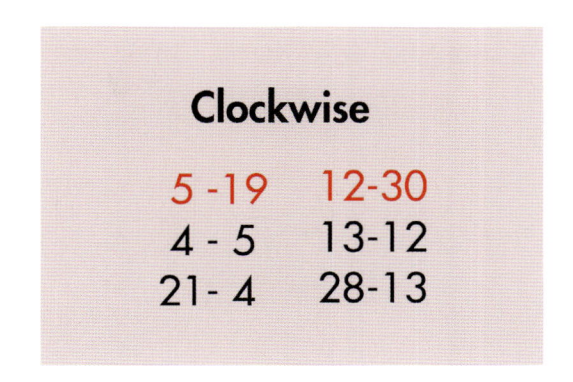

## Design 8.1 Spiral Braid with Variations

Design 8.1.1 is made with clockwise movements while Design 8.1.2 is made with anticlockwise movements, thus changing the slant of the spiral. In Design 8.1.3, a zigzag pattern is created by alternating clockwise and anticlockwise steps.

For both stand and card braiders, the pattern sequence of Design 8.1 is easy to learn if attention is paid to which one of a pair of threads is moved and where it is placed in relation to its new neighbors. With a little practice, card braiders will be able to follow the arrows shown on the stand diagrams instead of relying on the numbers. This takes less concentration and allows more attention to be given to the structure of the pattern.

Design 8.1.3 is a two-color zigzag made by alternating clockwise and anticlockwise steps that change the direction of the spiral. In order to do this *without* creating floats, (elongated stitches that appear as errors), the upper threads are lifted over the lower threads before changing direction as indicated by the curved arrows in figure 5.7. Step-by-step instructions are given on the Design page.

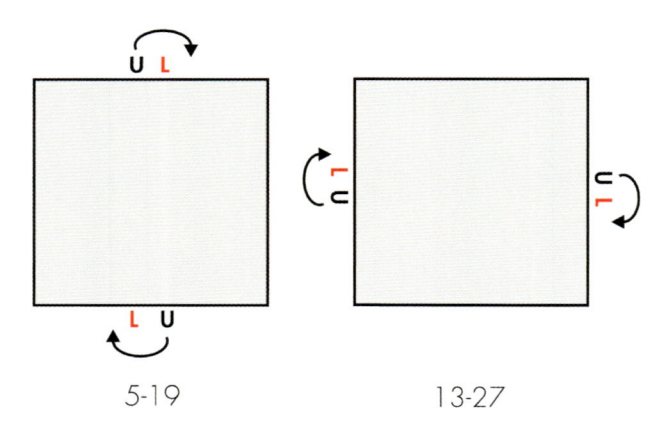

## Design 8.2 Round Braid

The three variations in Design 8.2 are made with alternating clockwise and anticlockwise steps developed from a simpler 4-strand braid structure.

8.2.1.
Two-color vertical stripe.

8.2.2.
Two-color chevron.

8.2.3.
Two-color staggered chevron.

5-19        13-27

**FIGURE 5.7.**
To avoid making unwanted floats when reversing the spiral in Design 8.1.3, the upper threads are moved to the other side of the lowers before making the move.

**FIGURE 5.8.**
Contemporary knitted purse with a 16-strand reversing spiral design used in the strap, similar to designs used in slings, made by Ilda Boza Canto.

62

## Design 16.1 Spiral Braids

Patterned 16-strand braids are an integral part of Andean sling braid design. Figure 5.8 shows a beautiful purse made by Ilda Boza Canto, from Yauli, in the central highlands of Peru. Her finely braided purse strap uses a reversing spiral pattern, similar to Design 16.1.4, combined with a stripe pattern. Other 16-strand braids frequently seen in slings include spiral, striped, chevron, diamond, and lattice patterns.

Designs 16.1.1 and 16.1.2 rely upon the clockwise and anticlockwise moves shown in figures 5.9 and 5.10. The zigzag pattern in Design 16.1.4 is made by continually reversing the structure after a given number of steps. For card braiding, the four inner slots on each side of the 24-strand card are used as the home slots leaving two slots empty on either side.

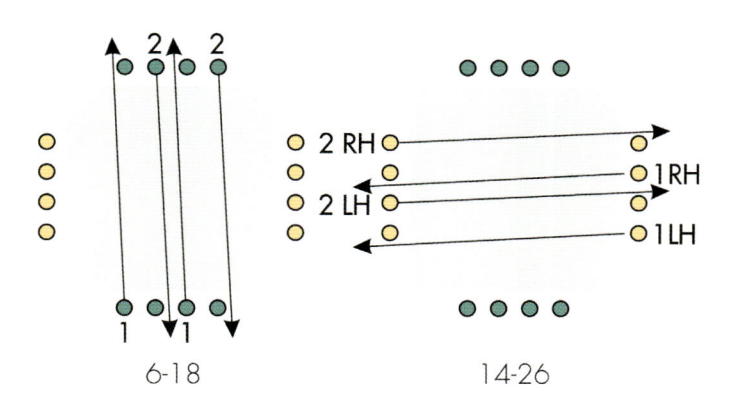

**FIGURE 5.9.**
Clockwise movements used in Design 16.1.

**16.1.1.**
Two-color clockwise spiral.

**16.1.2.**
Three-color anticlockwise spiral with thick and thin stripes.

**16.1.3.**
Four-color clockwise spiral.

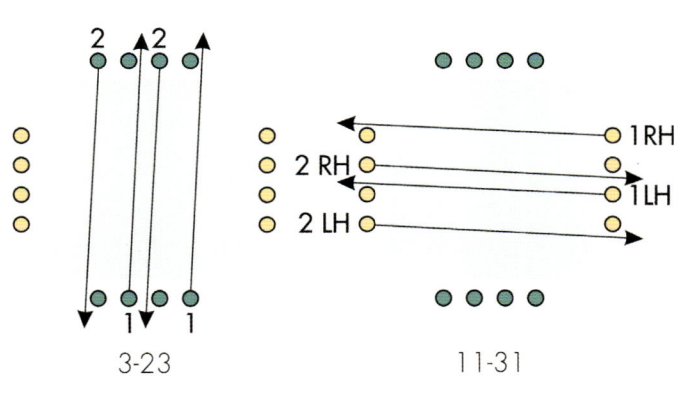

**FIGURE 5.10.**
Anticlockwise movements used in Design 16.1.

**16.1.4.**
Four-color reversing spiral.

## Design 16.2 Square Braid Variations

The square braids in Design 16.2 are made by repeating two steps: the N/S threads turn clockwise; the E/W threads turn anticlockwise. While the 8-strand version of the braid can be made with three variations, this 16-strand version has four with the potential for more by changing the colors or their positions. Some of the variations, like Designs 16.2.2 and 16.2.4, will have the colors reversed on the adjacent sides, giving them a more dynamic character when viewed from the corner of the braid.

**16.2.1.**
Two-color vertical stripe.

**16.2.2.**
Vertical stripe with diagonals.

**16.2.3.**
Two-color chevron.

**16.2.4.**
Two-color stripes and squares.

## Design 16.3 Diamonds and Zigzags

Design 16.3 introduces the *inversion move* which allows the creation of diamond patterns. It is used to change the structure of the braid by reversing the direction of the rotation. In an inversion move, two lower threads swap places, changing their positions to upper threads, allowing the rotation of the threads to reverse. This key structural change will be seen frequently in 24-strand sling braids. Diamond patterns are created by alternating rotation moves with inversion moves. (See figures 5.11 and 5.12.)

**16.3.1.**
Two-color diamonds.

**16.3.2.**
Three-color diamonds.

**16.3.3.**
Three-color diamonds.

**16.3.4.**
Two-color zigzag.

*Key moves for making inversion steps on the card.*

The notation for rotation and inversion moves for braids of 16-strands (or more) no longer appears as columns of numbers on the Design pages. Instead, *only the key moves will be shown below the gray boxes,* e.g., 6-18 in figure 5.11, and 22/4 – 20/6 in figure 5.12.

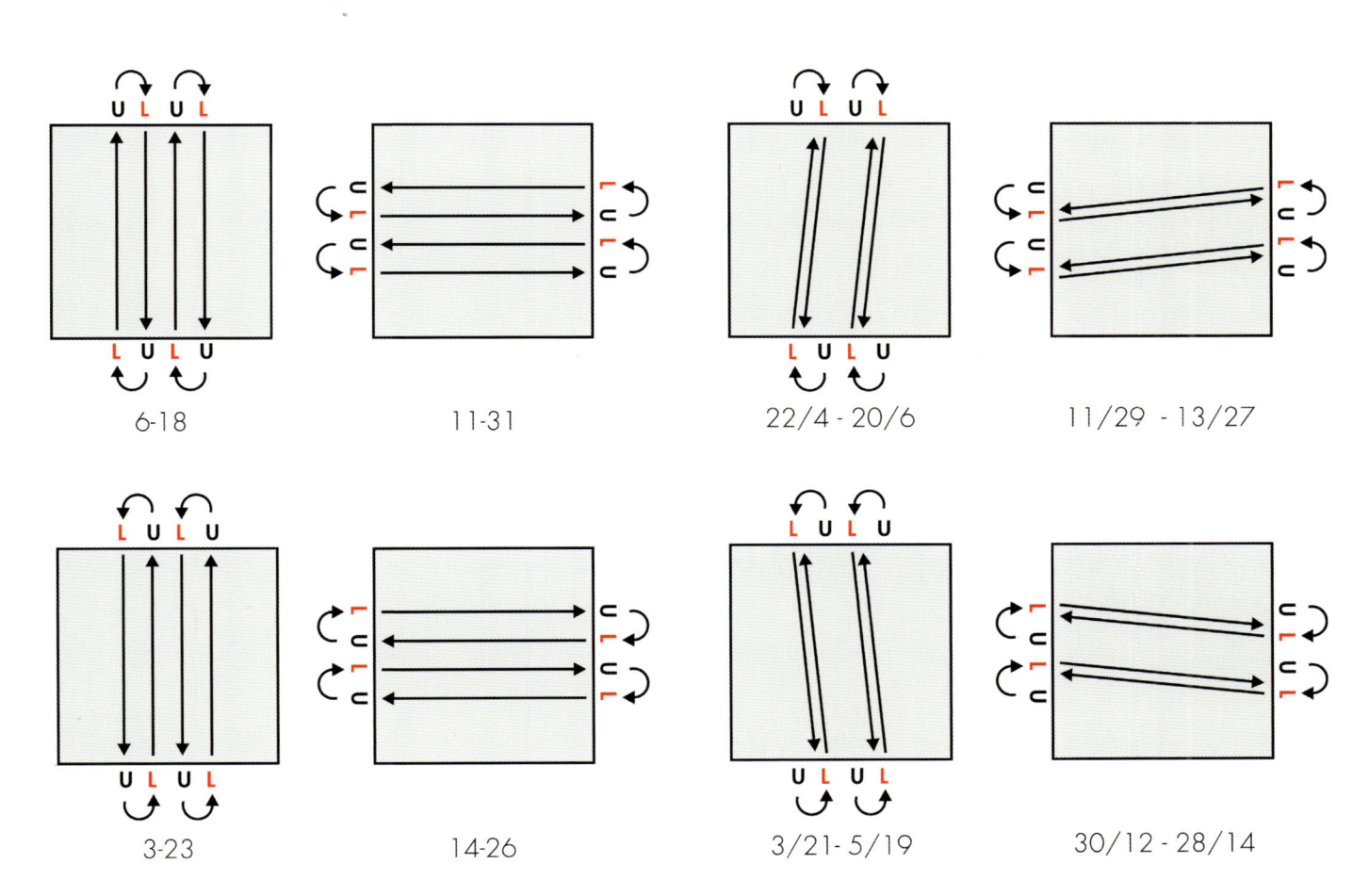

6-18

11-31

22/4 - 20/6

11/29 - 13/27

3-23

14-26

3/21 - 5/19

30/12 - 28/14

**FIGURE 5.11.**
Rotation moves with key moves for Design 16.3.
A dash between the numbers, e.g., 6-18, always indicates a rotation move where threads move to new slots.

**FIGURE 5.12.**
Inversion moves with key moves for Design 16.3.
A slash between numbers, e.g., 22/4, indicates that the threads trade places using the same slots.

65

**FIGURE 5.13.**
Andean sling with white chevrons and multicolored diamonds on a bi-colored ground. The cradle features a simple double diamond pattern with a wide slit.

# Chevron Designs 24.1–24.3

Andean core-carrying braids of 24-strands are complex structures, similar to those found in some woven bands and in double cloth pick-up braids for takadai, where designs on the surface are created by thread substitution. Cores range from 8–24 strands, allowing a wide variety of exciting design choices. The sling shown in figure 5.13 is an example of how simple patterns can be varied along the length of a braid. The designs in this book carry cores of 8–12 strands with one of 16 strands and eight designs that can be made without cores.

The basic 24-strand square braid used for core braids is a development of the 8-strand braid shown in figure 5.14. Design 8.1 *(left)* has two threads on each N, S, E, and W side. Design 16.1 *(center)* is an extension of this 8-strand braid with an additional pair of threads on each side. Design 24.1 *(right)* adds four more pairs resulting in a 24-strand braid. The red dotted lines on Designs 16.1 and 24.1 denote the boundaries of the 8-strand braids that join to make the more complex braids. The rotational movement shown by the pair of arrows in Design 8.1 is repeated with each successive set of threads in Designs 16.1 and 24.1.

Because of the way in which the threads rotate around a center point in square braids, Designs 8.2, 16.1, and 24.1–24.14 can carry cores. The square structure of 24-strand sling braids is created by the way that the N/S threads interlace with each other followed by the E/W sets of threads. They retain their square shape even when there is a set of threads in the core. These braids take additional time to make because attention must be paid to tensioning, however, the time is well worth it! Instructions for the basic tensioning techniques are described below.

**FIGURE 5.14.**
The 8-strand braid *(left)* is the "parent" of the 16-strand braid *(center)* and the 24-strand braid *(right)*. Dotted lines separate each 8-strand section. Notice that as additional sets of threads are added to the N/S and E/W of the composite braid, the rotational movement within each 8-strand section remains the same regardless of the color order.

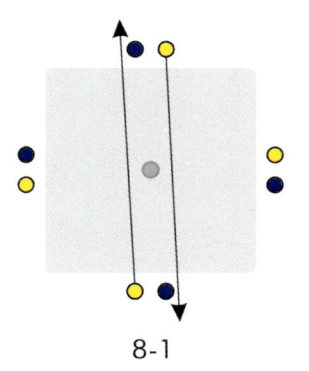

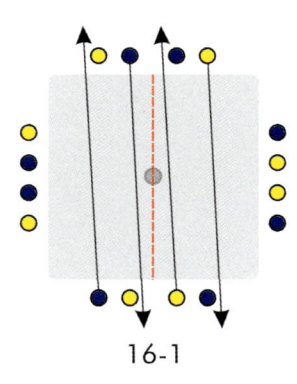

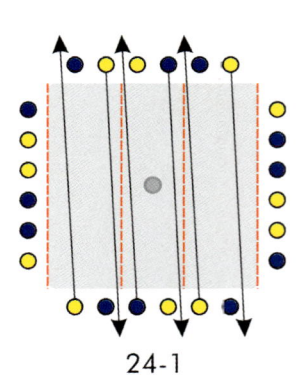

8-1    16-1    24-1

## Making 24-strand Chevron Patterns

Twenty-four–strand core braid patterns are derived from the basic structure of chevron Design 24.1. Most use the same color sequence of ABBAAB seen on the N side of Design 24.1 in figure 5.14. Notice that the color order is mirrored on each *adjacent* side of the braid so that at the beginning of each N/S step, the colors of the corner threads should match. Therefore, *opposite* sides of the braid will be in reverse color order. Recognizing the correct color arrangement is an essential skill that helps the braider quickly spot errors.

This design can be made with either a V- or an A-shaped pattern. Chevrons are made with repetitive clockwise/anticlockwise or anticlockwise/clockwise steps.

## Two Methods of Making 24-strand Square Braids on the Core Stand

Two methods are described in the Appendix for making 24-strand square braids on the stand in which the braid structure is the same but the sequence of thread movement differs. In Method 1, the braider stands and begins by moving pairs from the same side of the stand. Method 2 can be worked sitting or standing. In this case, the threads are moved from opposite sides of the stand, diagonally across from each other. (See Appendix C, figures A.8–11.)

A note about reading the diagrams in this chapter: Some braiders are very comfortable using diagrams and will be able to look at the visual instructions and see how to reproduce those motions on the stand. Many braiders find detailed written instructions essential. Both are included. It takes many words to describe what will seem simple and straightforward once your brain has come to its own understanding of the essential skills: how to identify the threads to be moved, and how to spot their destinations. We suggest rewriting the instructions using the spoken or spatial cues that are meaningful to you. It may also be helpful to use the blank design planning diagram, to record the color changes that occur as the braid is worked. (See Appendix D.)

## Working Tension for Card and Stand

Although tensioning methods differ for card and stand braiding, the goal is the same. Careful tensioning will create even stitches and a square shape to the braid.

*Tensioning for the card.*

When working on a card, the hands must do the work that the bobbins and counterbalance do when working on a stand. As with braids that have fewer threads, while moving the threads across the card to their new positions, the braid must be held under tension on the underside of the card. The tensioning can be controlled by holding the braid firmly between two fingers while the threads are moved. One way to do this is shown in figure 5.15. To help keep the braid square as it is worked, every few steps hold the braid between your thumb and finger close to the point of braiding and knead on all four sides. (See figure 5.16.)

**FIGURE 5.16.**
To keep the braid square, hold it close to the point of braiding and knead on all sides.

**FIGURE 5.15.**
Hold the braid underneath the card between the index and middle fingers to help tension the braid.

*Tensioning for stand braiding.*

As with most braids made on a stand, the braid is tensioned in part by the action of the counterbalance weight against the weight of the bobbins. *Before* working a step, identify the three pairs of opposite threads to be worked, tension each pair by pulling away from the point of braiding, then make the moves according to your method.

Additional tensioning instructions for diamond braids are covered in the next chapter.

*Counterbalance weight.*

Many factors are involved in choosing a counterbalance weight for core braids. There is no formula that works for all braids. Two essential factors are the desired stiffness or flexibility required of the finished braid, and the characteristics of the chosen yarn. All 24-strand braid samples in this book were made using 70 g bobbins. Eighty-five gram bobbins can also be used.

The counterbalance weights used to make the 4–16-strand samples are noted on their Design pages. The 24-strand designs do not specify a counterbalance weight. We suggest between 1100–1500 grams (39–53 oz.). It is highly recommended that you make samples with different counterbalance weights to determine if the braid has the desired amount of stiffness or flexibility. *Always check the stiffness with the braid off-tension by inserting a knitting needle into the braid on the underside of the stand and then sliding the counterbalance down the braid. Check the softness/stiffness of the braid and then adjust the counterbalance accordingly.* Increasing the counterbalance weight will create longer stitches and a looser, more flexible structure; decreasing the weight will yield a tighter braid.

## Checking for Errors

When working with a suspended core, the three most common errors are: picking up the wrong thread; placing a thread in the wrong position; or for stand braiding, accidentally knocking a thread out of position. The last is most likely to happen if the bobbins are at different heights and tangle as they are moved. Re-adjust the bobbin height regularly. To help reduce errors, check two things after completing each of the two-step sequences:

1. Compare the color order in the next diagram to be worked with what is on the stand; they should match.
2. Look at the new set of number #1 threads and check to see that they are in the lower position before making the move.

Shortly after completing the first eight or so steps, check the "squareness" of the braid, both by taking a look and by reaching under the stand and feeling the sides of the braid between your thumb and index finger. Sometimes a braid begins with a rectangular structure as a result of applying more tension to either the N/S or E/W sides of the braid as it was worked. If the shape is not correct, undo the braid and begin again.

*Using a lifeline.*

A "lifeline" can help both card and stand braiders manage errors. (See figure 5.17.) It can be helpful when there is a long pattern sequence in which there is a greater chance for errors. Should a mistake be made, you can find the beginning of the sequence by unbraiding to the lifeline and then starting over. The lifeline eliminates the stress of keeping track of steps as you work them backwards. Use a thin cotton thread similar to a #30 crochet cotton.

Cut a strand about 12" (30 cm) long. Insert one end of the lifeline before beginning the first step of a pattern as follows:

1. Place one end of the lifeline across the south threads of the braid as shown in figure 5.17. Do not tie; instead, keep the short and long ends tucked down alongside the braid. Complete the sequence of steps. Check for errors.
2. If the sequence is correct, pull the long end of the lifeline up to the top of the stand and place it across the south threads. Resume braiding.

If you find a mistake, work the steps backward until you find the error. Work back an additional step or two to be sure you have identified the next step. Check that uppers and lowers are in the correct positions before beginning again. When the braid is completed, remove the lifelines.

**FIGURE 5.17.**
A "lifeline" (the white cotton thread) has been looped over the south threads to mark the beginning of a pattern repeat.

## Design 24.1 Classic Chevron Braid in Two-, Three-, and Four-color

Design 24.1, made without a core, is a V-shaped chevron that is the bottom half of the diamond design. Using different colors for the N/S and E/W sides of the braid is a strategy that can be used with many other designs. The instructions are found in the Design pages that follow. Practice braiding this pattern so that you are very familiar with the basic 24-strand braiding movements before making Design 24.2 which details how core exchanges are made.

**24.1.1a.**
Two-color chevron.

**24.1.1b.**
Three-color chevrons.

## Design 24.2 Three-color Chevrons with Core Changes

The three patterns seen in Design 24.2 are made possible by core exchanges. Although the methods for the stand and the card are similar, the methods of tightening the braid differ and are explained separately. We recommend making a sample of Design 24.2, following the instructions below, before making the other designs in this chapter. Begin by making a warp that is 20" (50 cm) long. Use the colors shown or substitute (see *Color Substitution When Sampling*, below). Bind, cut, and set up the braid on the core stand or card as described in Chapter 4, *Preparing the Warp*.

### Color Substitution When Sampling

If substituting colors when making a design sample, choose colors of similar color value to those shown in the instructions. Retaining the dark/medium/light values of the original colors will make following the diagrams easier. For instance, in Design 24.2, the dark brown/orange/gold could be replaced with black/gray/white or navy blue/medium blue/pale green. You can use the blank design planning template, found in Appendix D, to record your experiments.

24.2.1a      24.2.1b      24.2.1c

**24.2.1**
Core changes can alter the background or foreground colors on all surfaces of a braid, i.e., brown pattern on an orange ground (24.2a), followed by brown pattern on a gold ground (24.2b), and ending with brown and gold pattern on an orange and gold ground (24.2c).

### Interpreting Core Exchange Diagrams

For stand braiders, the six core change steps used to alter the color of the chevrons for Design 24.2 are detailed in Appendix E. All designs will use similar sequences of making core changes unless otherwise noted on their design pages. Because braiding Methods 1 and 2 move the threads in different sequences, the order of working the threads during a core change will differ. *Card braiders will use the same process with the exception that there is no need to select core threads in a specific order.*

## Steps for Exchanging the Core on the Card

When working core exchanges on the card, the core is held to one side during the first half of the moves, then flipped to the opposite side as the second half of the sequence is completed. The asterisks seen in the columns of card move numbers indicate when the core needs to flip. (See figure 5.18.) The numbers in red and green in the column of moves indicate threads that are in the lower position that are available for core exchanges if the design calls for it.

The braid in figure 5.19 had gold threads on each side of the card before the exchanges began. It now shows a gold thread that was moved from slot 7 to the core, and a brown thread that is being moved to slot 17. The brown thread now covers the orange thread and will hold it in place after the braid is tightened.

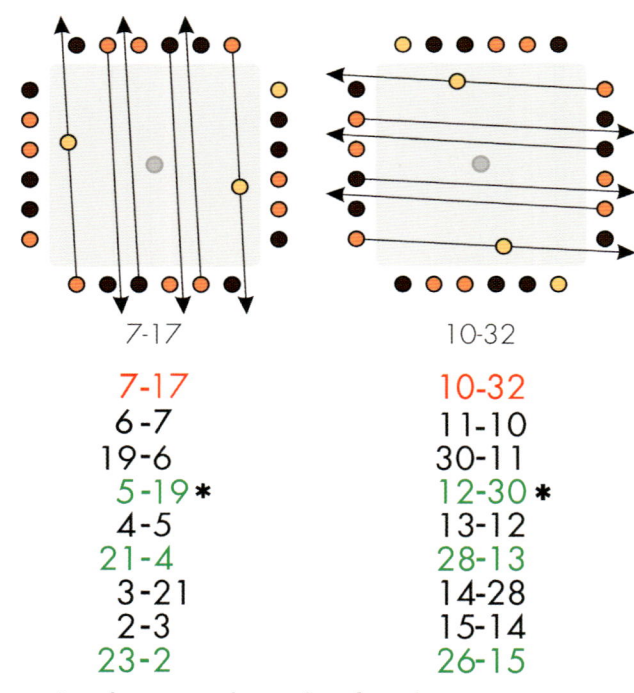

Figure 5.20 shows the result of the core changes. The gold threads have been moved to the core and have been replaced by brown threads. During the exchange, core threads were pulled one at a time to take up any slack.

7-17          10-32

7-17          10-32
6-7           11-10
19-6          30-11
5-19*         12-30*
4-5           13-12
21-4          28-13
3-21          14-28
2-3           15-14
23-2          26-15

\* Flip core threads after this move

**FIGURE 5.18.**
V-shaped chevron moves on the card.
Green and red numbers denote lower threads that are available for exchange with the core.

**FIGURE 5.19.**
The core is flipped out of the way while a set of moves are completed. A gold thread from the braid is exchanged with a brown thread from the core.

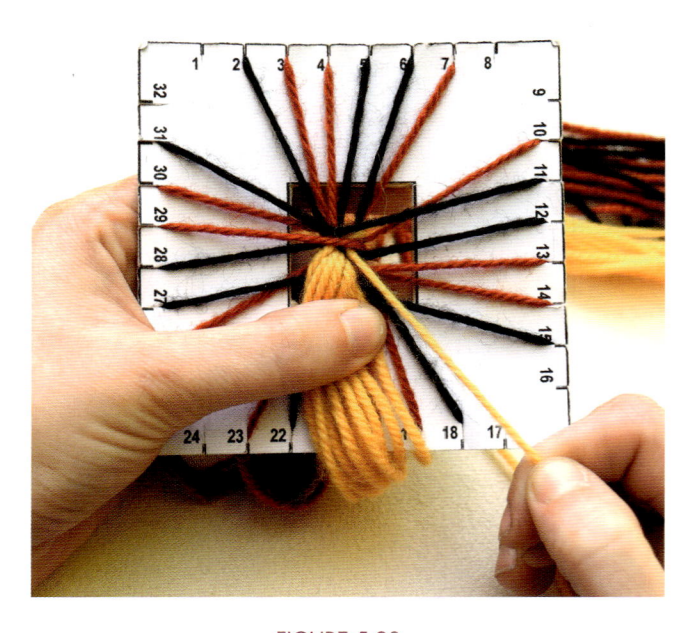

**FIGURE 5.20.**
Threads that have been moved to the core are pulled one at a time to take up any slack.

## Adjusting the Tension for Card Braids

As the braid grows, it will sometimes be necessary to adjust the upper strands on all four sides of the card to maintain the square shape. This corrects any variations in the tension that have occurred as the braid has been worked. The center four threads rarely need to be adjusted.

*The adjustment is made just prior to working a rotation move.* Figure 5.21 shows N/S pairs of upper threads in slots 3 and 23 on the W side of the card and in slots 7 and 19 on the E side. Hold the braid under the card with one hand, and with the other, pull the upper threads firmly away from the braid one at a time. Watch how the threads tighten. Judging how much tension is needed is a skill that comes with practice. While tensioning before a move seems counterintuitive, it snugs the lower threads from the previous step into position. It may be necessary to tighten the "lower" partner threads from time to time.

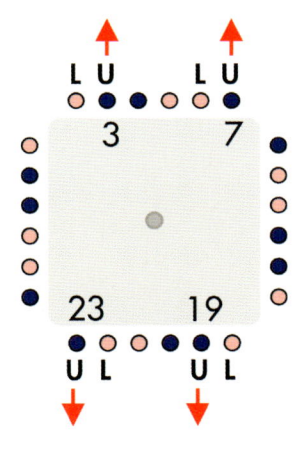

**FIGURE 5.21.**
Tension the braid where indicated prior to working the next step.

## Design 24.3 Chevron Braid with Vertical Stripes

Design 24.3.1, known as the *Palma* braid in Peru, is made without a core. It uses an alternate color arrangement in which pairs of yellow threads in the center of each face of the braid create the effect of "stitching." (See figure 5.22.) A description of how this design was used in a sling is found in Chapter 7, *Laverne Waddington's Story*.

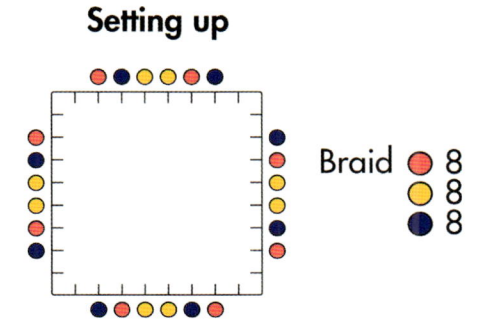

**Setting up**

**FIGURE 5.22.**
The Palma braid features an alternate color order in the set up.

**24.3.1.**
Chevron broken by a stripe on the face of the braid.

71

**FIGURE 5 23.**
Andean sling with diamond, chevron, and lizard motifs spaced along a dark-color ground. *Collection of K. Owen.*

## Designs 24.4 and 24.5: Making Diamond and Chevron Designs Using Inversion Moves

The thirteen designs presented in this section demonstrate how the variety of pattern possibilities expand as the techniques used in 16-strand braids are applied to 24-strand braids especially when a core is added. Core changing, combined with stitch inversion (reversing the direction of a pattern) allow diamonds of various styles to be added to fields of color. Patterns can also be created by altering the number of steps, or repeating steps. A variety of patterns can be spaced along the length of the braid as is often seen in Andean festival slings like the one shown in figure 5.23.

### Creating Patterns with Inversion Moves: Designs 24.5–24.13

*Inversion moves,* which were introduced in Design 16.3, allow the braider to make an important structural change to a braid by reversing the direction of the stitches thus allowing the creation of diamond and stripe patterns. The diamond and chevron pattern of Design 24.4.1 is created by alternating inversion moves with rotational moves. A variety of strategies for developing simple patterns are explored on its design page. The elongated chevron pattern in Design 24.5 is made with continuous inversion moves.

**24.4.1.**
Diamonds and chevrons created by combining rotation and inversion moves.

**24.5.1.**
Repeated inversion moves create a distinctive chevron with an elongated pattern.

**24.5.2.**
Elongated chevron with three possible color changes.

While inversion moves are customarily used to create diamonds, Design 24.5 uses repeated inversion moves to create a distinctive chevron with an elongated pattern. It can be made with or without a core and as either a two- or three-color braid.

## Walk-through of Making Single Diamond Pattern Design 24.6.1

The following walk-through describes making the single diamond motifs in Design 24.6.1. In this design, core changes and inversion moves allow a small diamond pattern to be added to a solid-color braid. We suggest making a sample following the instructions below before making the other designs in the chapter. Begin by preparing an 18" (45 cm) long warp with 24 ends of a dark color and 12 of a light color. Bind the ends for a tassel start. Set up the warp on the braiding stand or card in the color order shown in figure 5.24.

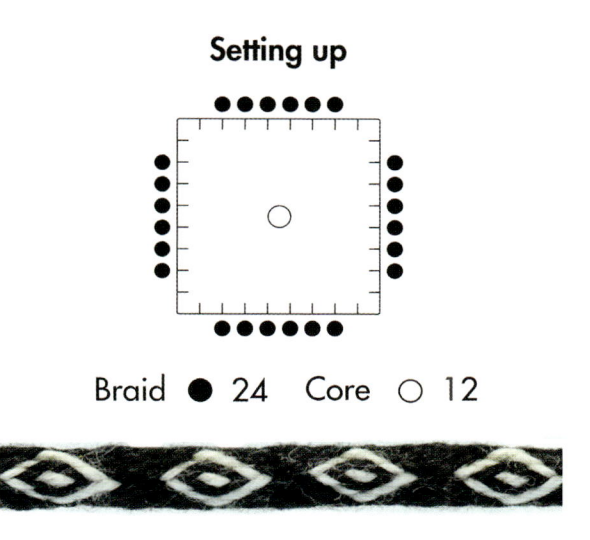

**Setting up**

Braid ● 24   Core ○ 12

**FIGURE 5.24.**
Design 24.6.1 begins as a solid-color braid with a dark core.

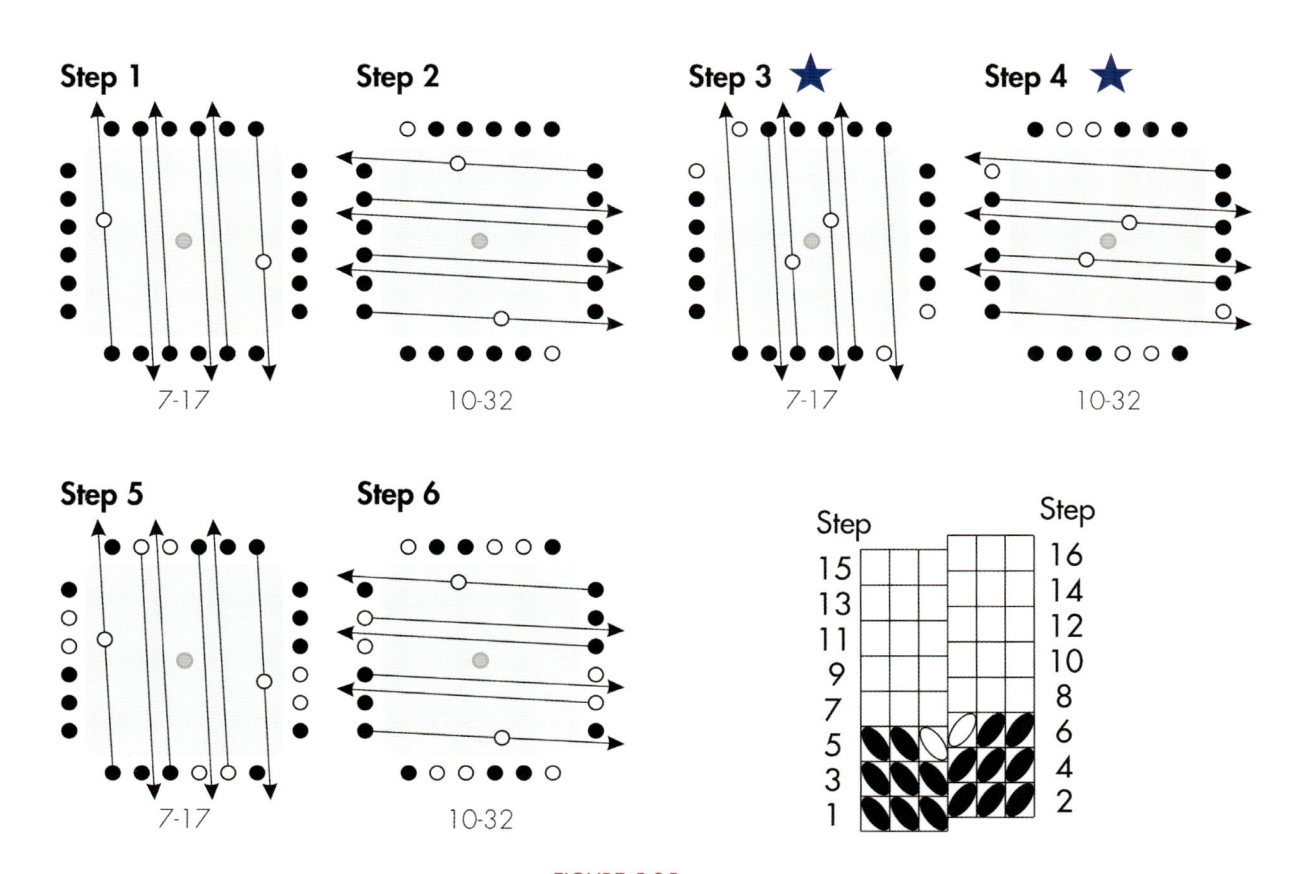

**Step 1** 7-17
**Step 2** 10-32
**Step 3** ★ 7-17
**Step 4** ★ 10-32
**Step 5** 7-17
**Step 6** 10-32

**FIGURE 5.25.**
Steps 1–6: The first six steps create the V-shape of the diamond pattern.

73

Figures 5.25–5.30 show the basic sequence for swapping the core threads in and out to create a diamond pattern. Steps 1–6 bring the light threads out of the core to make the lower "V" portion of the diamond. *Inversion Steps* 7 and 8 alter the structure of the braid by changing the slant of the stitches. The braid must be adjusted in Steps 11 and 12 in order to pull the inversion stitches into a vertical position. The last steps complete the diamond and return the light stitches to the core. Remember that there is a two-step delay between a core thread being exchanged and its appearance on the braid.

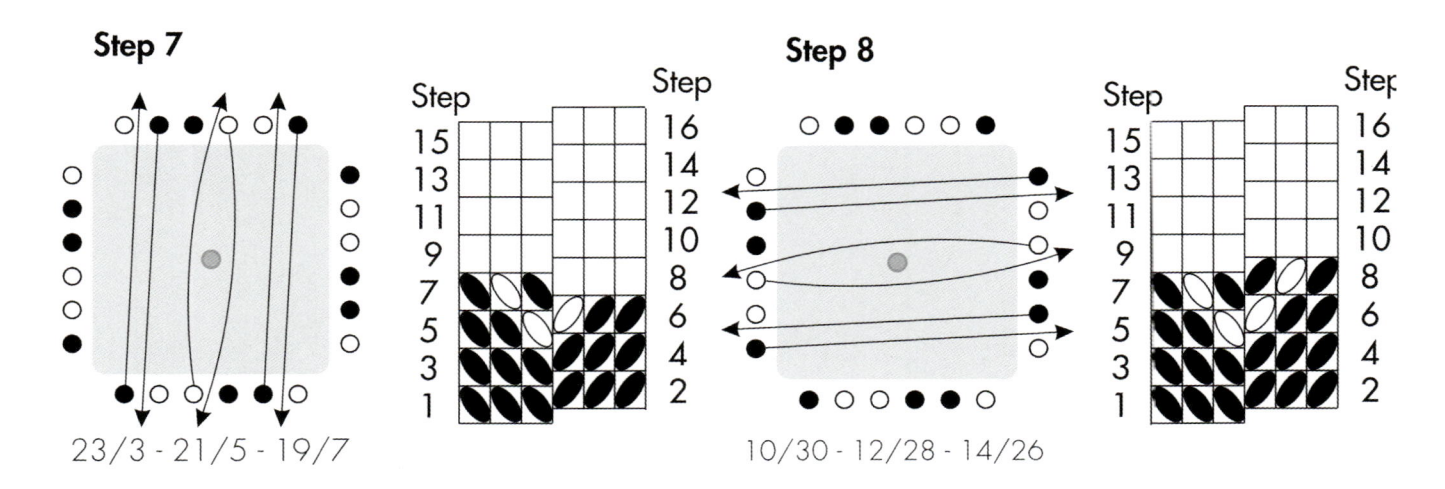

**Step 7**

23/3 - 21/5 - 19/7

**Step 8**

10/30 - 12/28 - 14/26

FIGURE 5.26.
Steps 7 and 8: Each pair of threads trade places between partners in this inversion step in order to reverse the position of the uppers and lowers. The swapped stitches will lie vertically and subsequent steps can now build the A-shaped portion of the structure to close the diamond. Stitches exchanged in previous steps become visible in the braid.

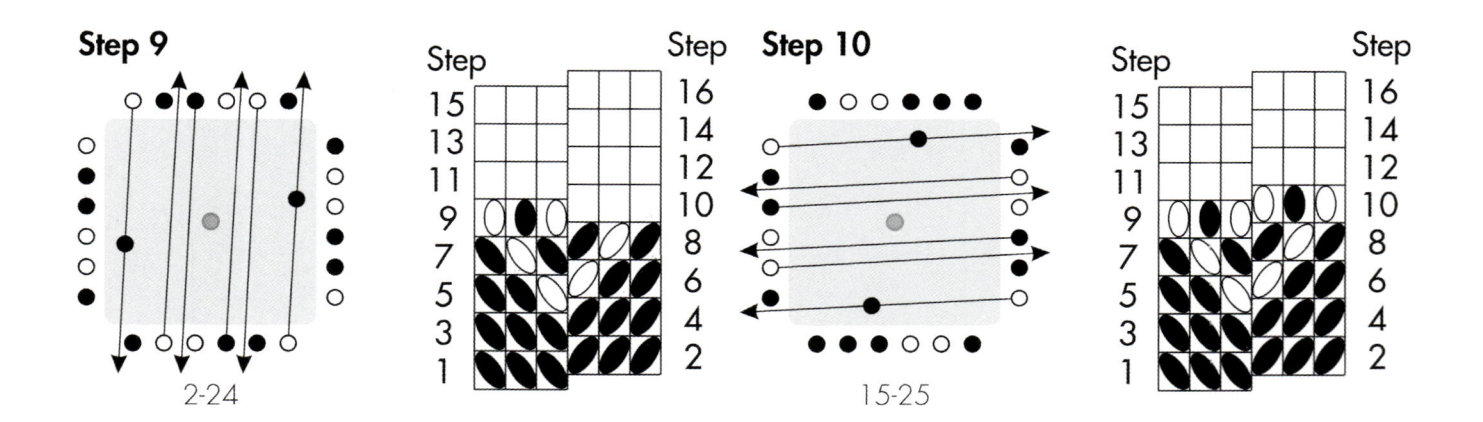

**Step 9**

2-24

**Step 10**

15-25

FIGURE 5.27.
Steps 9 and 10: These moves create the last stitches of the diamond. The white vertical stitches created in Steps 5 and 6 appear in rows 9 and 10 on the placement grids.

## Adjusting Tension for the Stand and the Card after Inversion Steps (or Mind the Gap)

**FIGURE 5.28.**
Diamond patterns require a tensioning step to prevent a break in the pattern as shown in the circled area.

When changing the shape of a diamond from a V- to an A-shape, it is necessary to make a tensioning adjustment to prevent a gap (a "grinning stitch") like the one that breaks the pattern in figure 5.28. The adjustment will ease the stitches created by the inversion steps into the correct vertical position. A *blue star*, at the top left corner of step diagrams, as seen in figure 5.30, indicates that tensioning is required *before* a step is worked. (Please note that it takes many words to describe something that is actually quick to do with a little practice.) After working the "V" portion of the diamond in Steps 1–6:

1. Work the inversion moves (Steps 7 and 8).
2. Work the next pair of rotational moves (Steps 9 and 10).
3. *Before* working the *A* portion of the diamond (which begins with N/S Step 11, figure 5.30), tension the braid using one of the following techniques:

    a. Stand: Pull each pair of circled threads firmly away from the braid in the direction of the red arrows. (See figures 5.29 and 5.30.)

    b. Stand or Card: Hold the braid under the stand or card with one hand and pull each of the four indicated threads against the point of braiding.

4. Work N/S Step 11 shown in figure 5.30.
5. Now tension the upper pairs of E/W threads as indicated by the circles and arrows.
6. Work E/W Step 12.

For both stand and card methods, take care to apply equal tension to the threads as they are adjusted in order to maintain the square structure of the braid. Note that tensioning the N/S threads adjusts the inversion stitches on the E/W side of the braid and vice versa.

Initially, this additional tensioning process may slow down the rhythm of braiding. After a while, it becomes part of the routine: make the N/S and E/W inversion steps; work N/S and E/W rotational moves in the new direction; tension N/S and work the step; tension E/W and work the step; (relax and) braid as usual to the turning point of the next diamond.

Depending on the yarns chosen, as well as the counterbalance weight, and the way each individual handles the threads, it may also be necessary to make the same adjustment to the partnered lower threads. (The center four threads rarely need to be adjusted.)

**FIGURE 5.29.**
To adjust the tension, hold and tighten the pairs of threads indicated by the circles and arrows in figure 5.30.

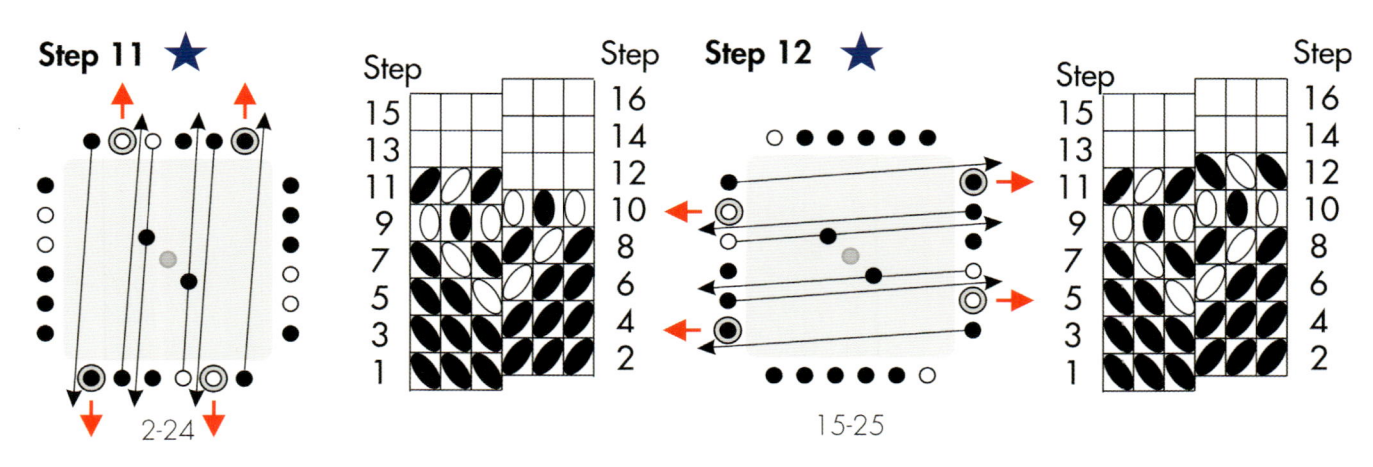

**FIGURE 5.30.**
Steps 11 and 12: The blue star indicates that tension adjustment is necessary prior to working a step. The circled bobbins indicate which threads must be tightened. the red arrows indicate the direction the threads are pulled.
See Design 24.6.1 for Steps 13–16 that complete the pattern.

# Designs 24.6–24.10. Two-color Diamond Variations

**24.6.1.**
Small diamonds on a plain-color ground.

**24.7.1B.**
Medium-sized diamonds on a dark ground.

**24.6.2.**
Continuous small diamonds made by repeating steps but omitting core changes that would return the braid to a solid background.

**24.7.1C.**
Diamonds display additional patterns when viewed from the side.

**24.8.1.**
Pattern created by working continuous diamonds for several repeats and then returning to the background color.

**24.8.3.**
Three-color variation with key pattern

**24.9.1.**
Linked diamonds.

**24.9.2.**
Continuous linked diamonds.

**24.10.1.**
Wave and spot pattern created by alternating rotation and inversion moves.

**24.10.2.**
Diamond chain pattern created by adding a third color to Design 24.10.1.

# Design 24.11: Alternate Method for Creating Diamonds

Design 24.11 uses both core changes and *thread exchange,* a new movement that is diagrammed as sets of curved arrows on the sides opposing the rotation or inversion moves. (See pp. 117–118 for Design 24.11.) The thread exchange is used to change the color order without using the core. It also alters the shape of this diamond by elongating it. When a step calls for a thread exchange, work the exchange first (between upper threads), then work the rotational or inversion moves. Thread exchanges are also used in Design 24.15.

**24.11.1A.**
Large open diamond with dark pattern on light ground.

**24.11.1B.**
Light pattern on dark ground.

# Multicolor Diamond Designs 24.12–32.1

Design 24.12. Braid Architecture Explained: Changing between Diamond and Chevron Patterns

This section presents a different way to view braid design. The instructions for Design 24.12 are presented in two ways. The 16 steps used to make the alternating red/white and black/white diamond pattern in Design 24.12.2 is detailed on the design page. But in addition, Table 5.1 is a resource that explains how to recombine those steps to create six other patterns. (See page 121.) Also important, *it gives the sequence of steps needed to transition from one pattern to another*, i.e., 24.12.1 (red/white chevrons), to 24.12.2 (alternating red/white and black/white diamonds).

Braiders can use the strategies detailed in Table 5.1 for transitioning from one design to another and create combinations not shown, i.e., red/black chevrons transitioning to red/black diamonds, a little diamond chain with a black outline and red eyes . . . . As you begin to see how the structures work, you will be able to create your own combinations. The yarn used for Design 24.12 is Persian yarn with two of the three strands per bobbin. This gives a slightly hairy appearance to the braids but the yarn has nice luster and makes braids that are hard wearing.

**24.12.1.**
Chevron braid with a V-fell that can also be changed to black/white, or black/red.

**24.12.2.**
Alternating red/white and white/black diamonds.

**24.12.3a.**
Red/white diamonds.

**FIGURE 5.31.**
Length of Design 24.12 with multiple pattern transitions like those found in headbands and Chumpita del Soltera braids.

78

**24.12.3b.**
Black/white diamonds.

**24.12.7.**
Three-color diamonds.

**24.12.4.**
Red/white diamonds with black eyes.

**24.13.1a.**
Thin threads were plied to create strands of varying thicknesses.

**24.12.5.**
Three-color little diamond chain.

**24.13.1c.**
Using yarns of the same thickness strengthens the outline of the brown strands but eliminates the embossed texture.

## Design 24.13 Embossed Diamonds with Thick and Thin Yarns

Design 24.13 is related to the Nasca braid in figure 1.28. Yarns of differing thicknesses were used to create an embossed pattern. (See *An Ancient and Contemporary Yarn Comparison* in Chapter 3.) Another variation, similar to Design 24.13, is seen in the contemporary Andean burden strap shown in figure 5.32, made with thick and thin strands. The color of one ring of the concentric diamond pattern alternates between tan (color A), gray (color B), and brown (color C). These are arranged with color A on the N/S and color B on the E/W axes of the braid. Later color B is replaced with color C. This strategy can be applied to other diamond patterns. (See Design 24.8.3.)

This continuous length of thick braid was probably used as a burden strap. The braid measures 5 yards × ¾" diameter (4.6 m × 2 cm diameter) and weighs 1.5 lbs. (680 grams).

**FIGURE 5.32.**
Contemporary Andean burden strap made with a pattern similar to
Design 24.13 but with bicolored diamonds surrounding the eyes.
*Collection of R. Owen.*

## Design 24.14. Four-color Diamonds with Cores of 16 Strands

This braid uses a two-color core of 16 strands. Four- and five-color diamonds are possible. The thicker core requires extra care when tensioning the braid. When the core increases to 16 strands, the color arrangement on the harness must be changed. (See Design page 24.14.)

**24.14.1a.**
Four-color diamond with a core of 16 strands.

**24.14.1b.**
Variation with colors of greater saturation and value contrast.

## Design 24.15 Three-color Diamonds Made without Cores

Design 24.15.1 does not carry a core. The pattern variations in this braid are created by repositioning four upper threads before each step. The focus of the pattern is the eye color that can be changed to make a braid with all dark, medium, or light "eyes." The braid can also be made with a variety of eye colors as seen in the handsome sling in figure 5.33

**24.15.1a.**
Three-color diamond pattern with alternating eye colors.

**24.15.1b.**
Three-color diamonds with dark outlines and eyes.

**24.15.1c.**
Variation with yellow outlines and eyes.

**24.15.1d.**
Variation with red outlines and eyes.

**FIGURE 5.33.**
Andean sling made with Design 24.15, a three-color braid made without core changes. The eye colors alternate. *Collection of R. Owen.*

Design 32.1 Textured Four-color Diamonds Made without Cores

Design 32.1 is made with 32 strands and no core. For those using the card method, a larger size card with 40 slots will be needed. (See Appendix D for details.) Color changes are made by repositioning upper threads. Figure 5.34 features an Andean sling braided with handspun alpaca in a lovely selection of natural colors.

**32.1.1a.**
Four-color elongated diamond pattern made without a core.

**32.1.1b.**
Elongated diamonds with lattice.

FIGURE 5.34.
Andean sling made with Design 32.1 using a variety of naturally colored alpaca yarns. *Collection of R. Owen.*

# THE DESIGNS

# 8.1
## 8-strand Spiral Braid with Four Variations

Made with c/balance 9.5 oz (270 grams)

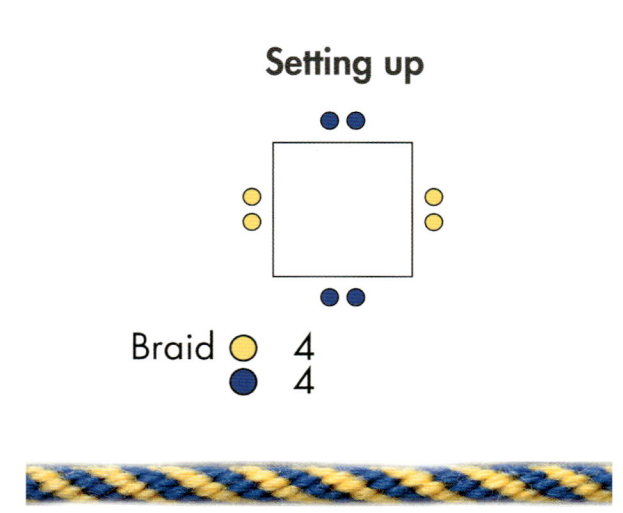

**Setting up**

Braid ○ 4
○ 4

### 8.1.1 Two-color clockwise spiral.
Repeat Steps 1 and 2, shown in Design 8.1.1 working instructions.

**Setting up**

Braid ○ 4
○ 4

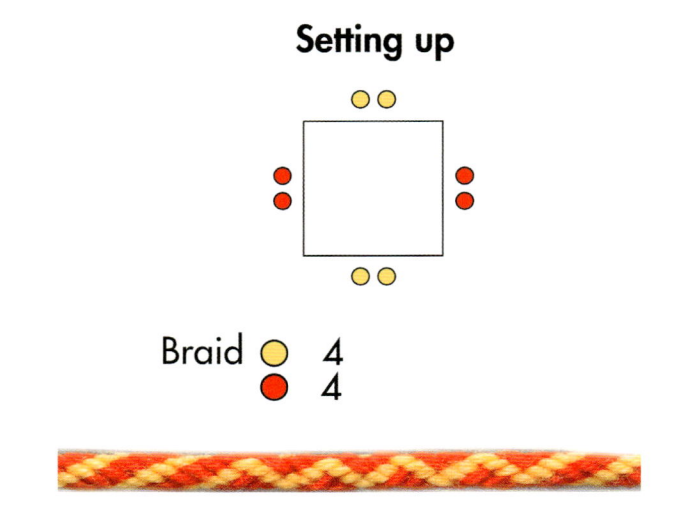

### 8.1.3 Two-color zigzags and reversing spirals.
The short zigzag pattern was made by changing direction every six steps. See the Design 8.1.3 working instructions.

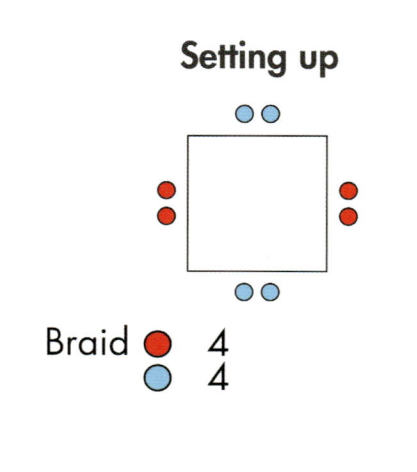

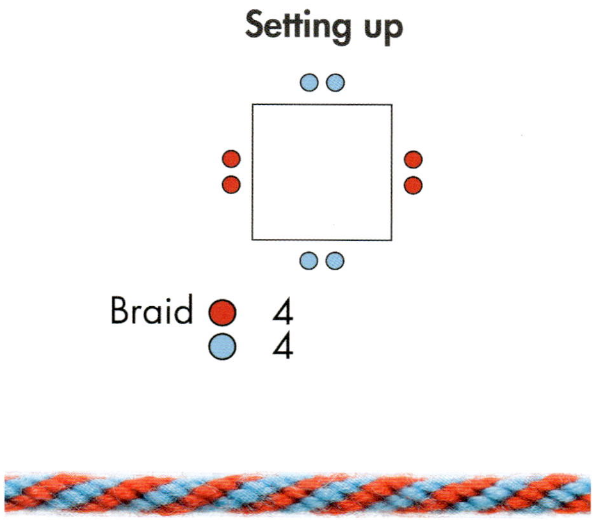

**Setting up**

Braid ○ 4
○ 4

### 8.1.2 Two-color anticlockwise spiral.
Repeat Steps 3 and 4, shown in Design 8.1.2 working instructions.

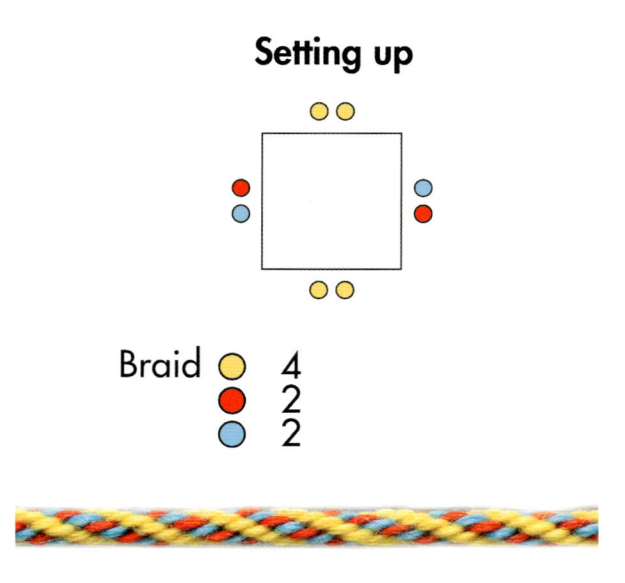

**Setting up**

Braid ○ 4
○ 2
○ 2

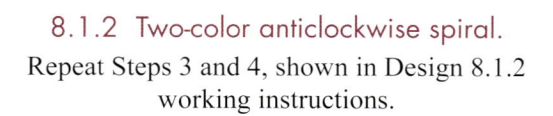

### 8.1.4 Three-color spiral.
Repeat Steps 1 and 2, shown in Design 8.1.1 working instructions.

*Working Instructions for Designs 8.1.1 to 8.1.4.*

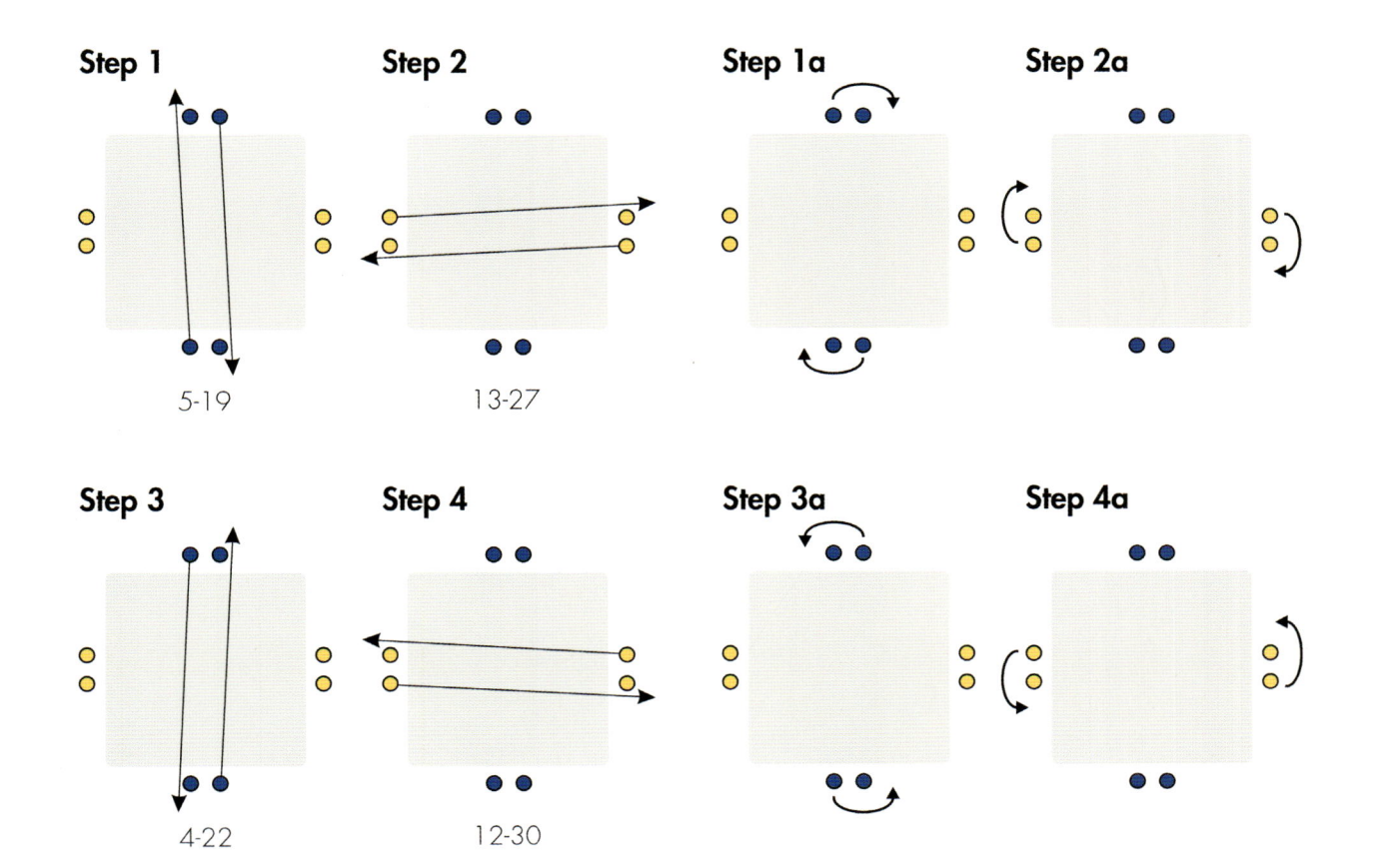

**Step 1**
5-19

**Step 2**
13-27

**Step 1a**

**Step 2a**

**Step 3**
4-22

**Step 4**
12-30

**Step 3a**

**Step 4a**

Design 8.1.1 and 8.4.

To make an 'S' slanting spiral, repeat Steps 1 and 2 for the desired length.

Design 8.1.2.

To make a 'Z' slanting spiral, repeat Steps 3 and 4 for the desired length.

Design 8.1.3.

To make the zigzag pattern:

1. Repeat Steps 1 and 2 three times. (The threads will be moving in aclockwise direction.)
2. Reverse the spiral to a 'Z' slant by first changing the position of the threads as shown in Steps 1a and 2a. Then repeat Steps 3 and 4 three times (anticlockwise).
3. To change back to an 'S' spiral, work steps 3a and 4a and then repeat Steps 1 and 2.

**Card Moves for Design 8.1.3**

| Clockwise | | Anticlockwise | |
|---|---|---|---|
| 5-19 | 13-27 | 4-22 | 12-30 |
| 4-5 | 12-13 | 5-4 | 13-12 |
| 21-4 | 29-12 | 20-5 | 28-13 |

On completing each column, reposition the threads to their set-up slots.

# 8.2
## 8-strand Vertical Stripe with Three Variations

*Made with c/balance 9.5 oz (270 grams)*

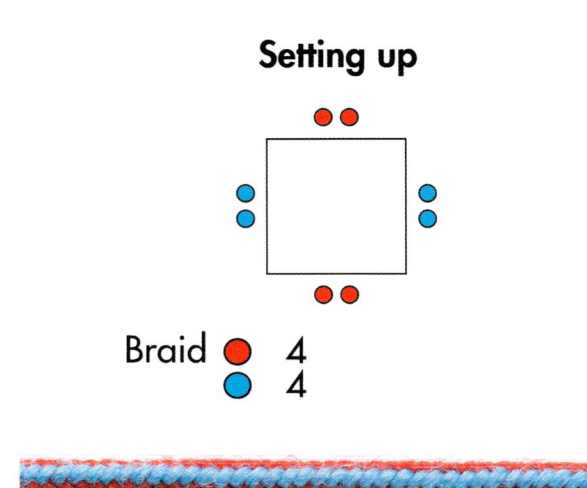

**Setting up**

Braid 🔴 4
🔵 4

**8.2.1 Two-color vertical stripe.**
Repeat Steps 1 and 2.

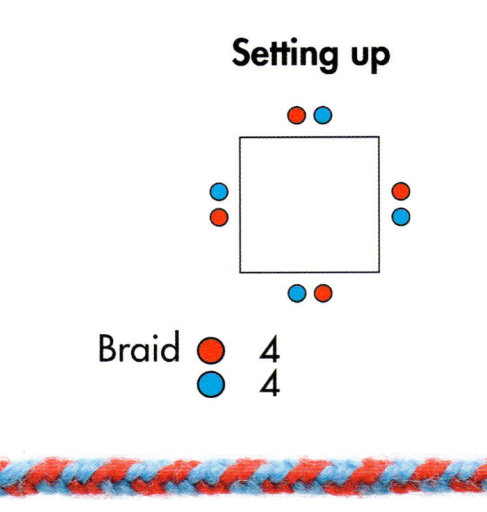

**Setting up**

Braid 🔴 4
🔵 4

**8.2.3 Two-color staggered chevron.**
Repeat Steps 1 and 2.

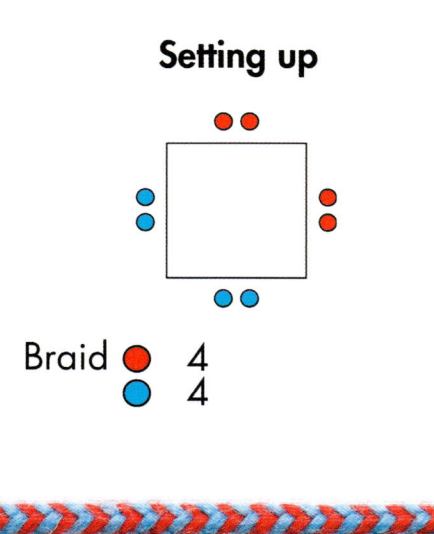

**Setting up**

Braid 🔴 4
🔵 4

**8.2.2 Two-color chevron.**
Repeat Steps 1 and 2.

*Working Instructions for Design 8.2*

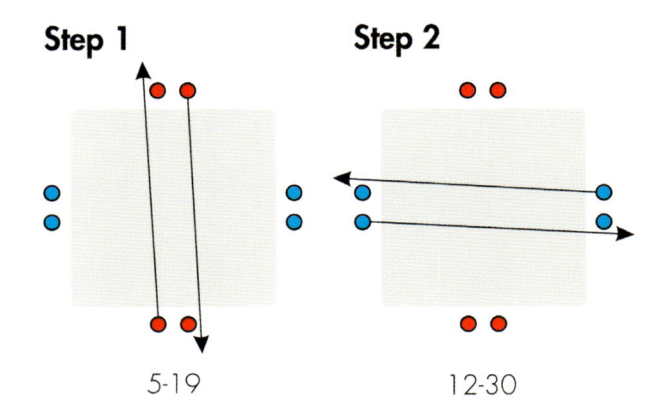

**Step 1**

5-19

**Step 2**

12-30

## Card Moves

| Clockwise | Anticlockwise |
|-----------|---------------|
| 5-19 | 12-30 |
| 4-5 | 13-12 |
| 21-4 | 28-13 |

On completing each column, reposition the threads to their set-up slots.

# 16.1
## 16-strand Spiral Braid with Four Variations

Made with c/balance 20 oz (570 grams)

**Setting up**

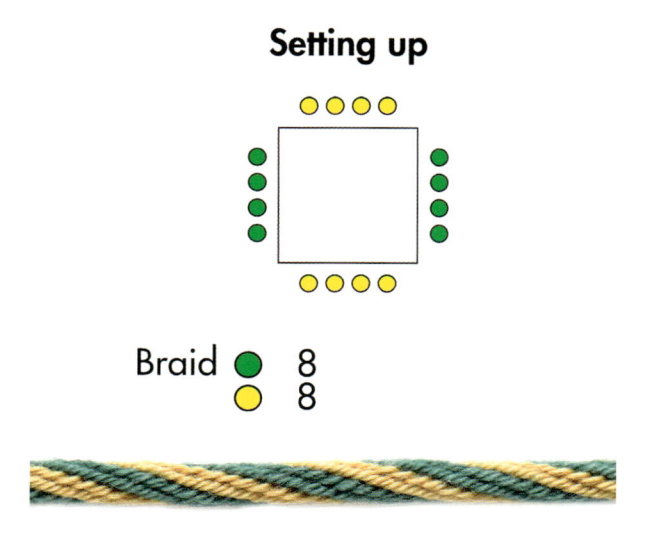

Braid 🟢 8
🟡 8

**16.1.1** Two-color clockwise spiral.
Repeat Steps 1 and 2.

**Setting up**

Braid 🔴 4 🟢 4
🟡 4 🟣 4

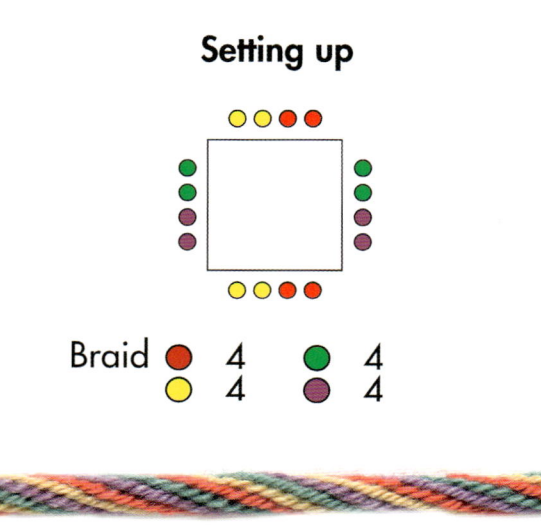

**16.1.3** Four-color clockwise spiral.
Repeat Steps 1 and 2.

**Setting up**

Braid 🔴 8 🟡 4
🟢 4

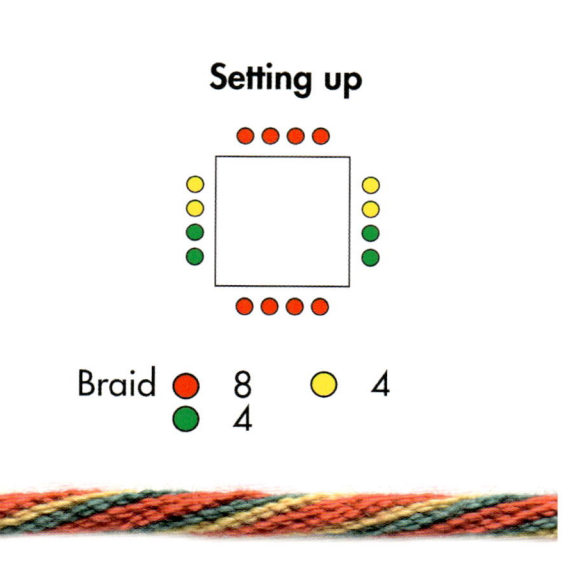

**16.1.2** Three-color anticlockwise spiral.
Repeat Steps 3 and 4.

**Setting up**

Braid 🟠 4 🟢 4
🟣 4 ⚪ 4

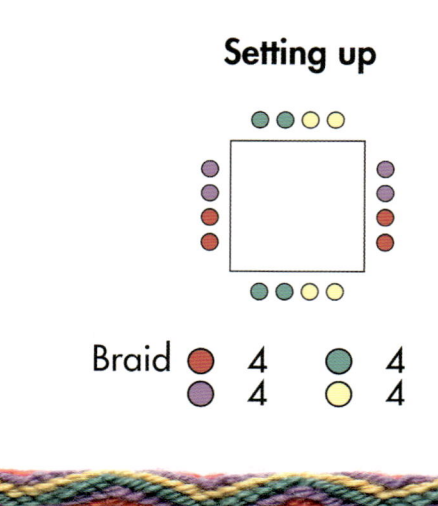

**16.1.4** Four-color zigzags.
Instructions for reversing the direction of the spiral are found on the following page.

*Working Instructions for Design 16.1*

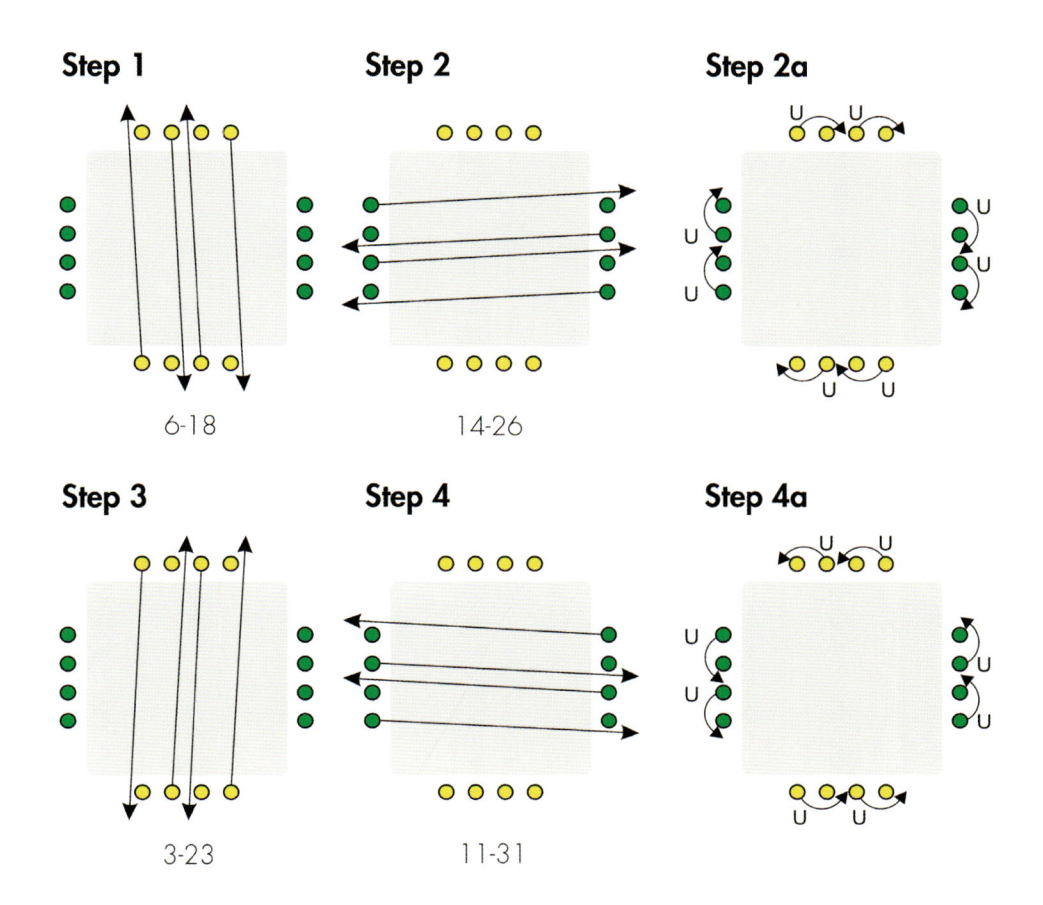

### Step 1
6-18

### Step 2
14-26

### Step 2a

### Step 3
3-23

### Step 4
11-31

### Step 4a

## To make a balanced zigzag pattern:

1. Work 12 repeats of Steps 1 and 2 ('S' spiral).
2. Reverse the spiral to a 'Z' slant by lifting the upper threads over the lowers, as shown in Step 2a.
3. Adjust the alignment of the threads and work Steps 3 and 4 for 12 repeats.
4. To revert to an 'S' spiral, reposition the threads by lifting the uppers over the lowers as shown in Step 4a. Continue with Steps 1 and 2.

## Card Moves

| Clockwise | | Anticlockwise | |
|---|---|---|---|
| 6-18 | 14-26 | 3-23 | 11-31 |
| 5-6 | 13-14 | 4-3 | 12-11 |
| 20-5 | 28-13 | 21-4 | 29-12 |
| 4-20 | 12-28 | 5-21 | 13-29 |
| 3-4 | 11-12 | 6-5 | 14-13 |
| 22-3 | 30-11 | 19-6 | 27-14 |

On completing each column, reposition the threads to their set-up slots.

# 16.2
## 16-strand Vertical Stripe with Eight Variations
*Made with c/balance 20 oz (570 grams)*

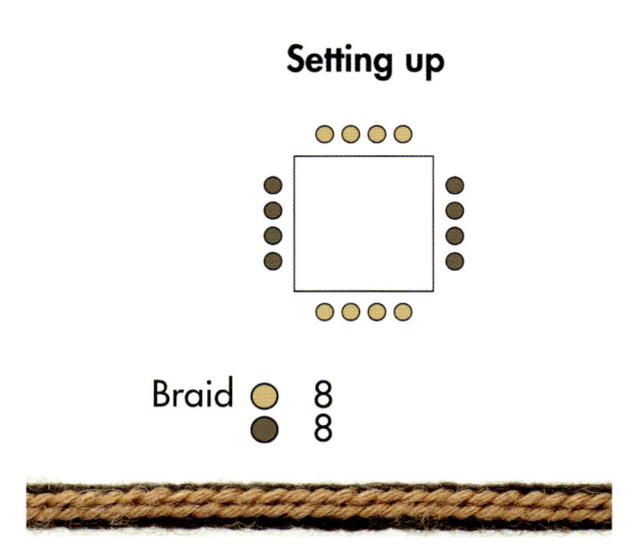

**Setting up**

Braid ● 8
● 8

**16.2.1** Two-color vertical stripe.
Repeat Steps 1 and 2.

**Setting up**

Braid ● 8
○ 8

**16.2.2** Two-color diagonal stripe.
Repeat Steps 1 and 2.

**Setting up for three- and four-color vertical stripes.***

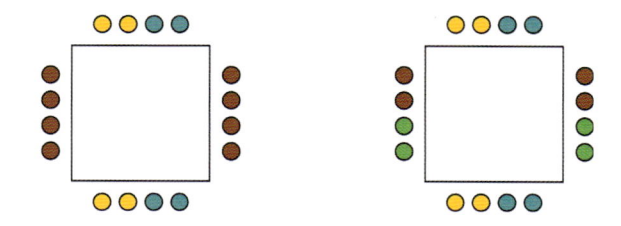

**Setting up for three- and four-color diagonal stripes.***

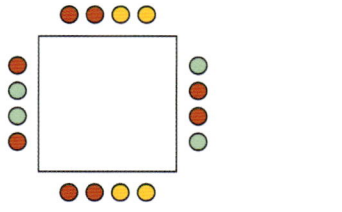

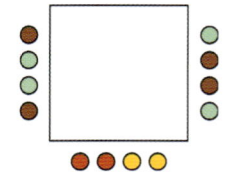

\* Braid samples are not shown.

\* Braid samples are not shown.

*Design 16.2 continued.*

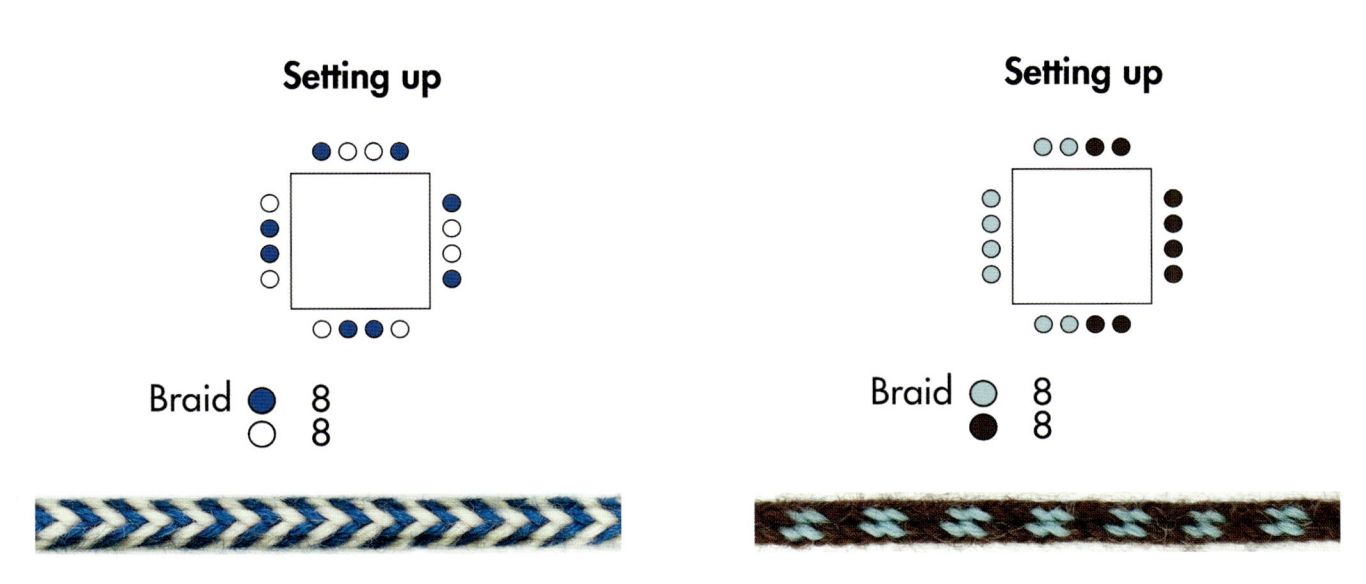

## Setting up

Braid ● 8
○ 8

**16.2.3 Two-color chevron.**
Repeat Steps 1 and 2.

## Setting up

Braid ○ 8
● 8

**16.2.4 Two-color stripes and squares.**
Repeat Steps 1 and 2.

---

*Working Instructions for Design 16.2*

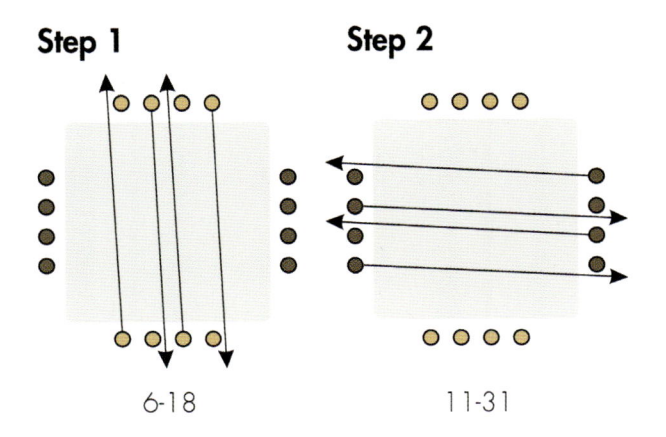

**Step 1**

6-18

**Step 2**

11-31

## Card Moves

Clockwise  Anticlockwise

| 6-18 | 11-31 |
| 5-6 | 12-11 |
| 20-5 | 29-12 |
| 4-20 | 13-29 |
| 3-4 | 14-13 |
| 22-3 | 27-14 |

On completing each column, reposition the threads to their set-up slots.

# 16.3
## 16-strand Diamonds with Four Variations

*Made with c/balance 20 oz (570 grams)*

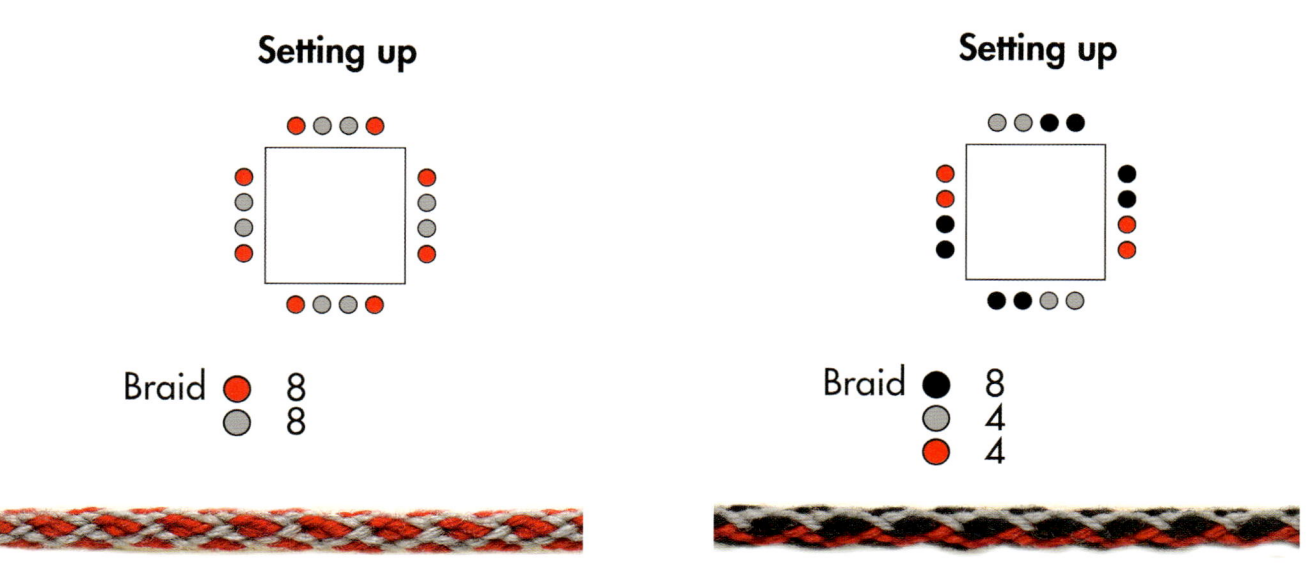

**Setting up**

Braid ● 8
○ 8

**16.3.1 Two-color diamonds.**
Repeat Steps 1–8.

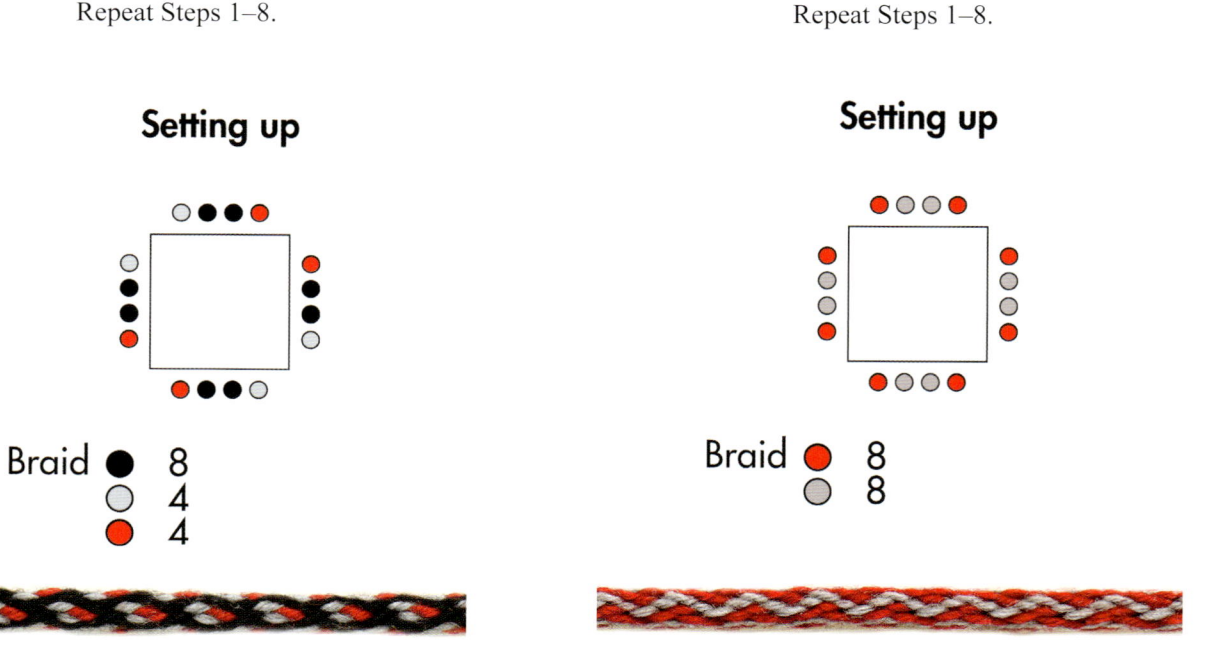

**Setting up**

Braid ● 8
○ 4
● 4

**16.3.3 Three-color diamonds.**
Repeat Steps 1–8.

**Setting up**

Braid ● 8
○ 4
● 4

**16.3.2 Three-color diamonds.**
Repeat Steps 1–8.

**Setting up**

Braid ● 8
○ 8

**16.3.4 Two-color wavy lines.**
Repeat Steps 1–8.

*Working Instructions for Design 16.3.*
Note that Steps 3, 4, 7, and 8 are inversion moves.

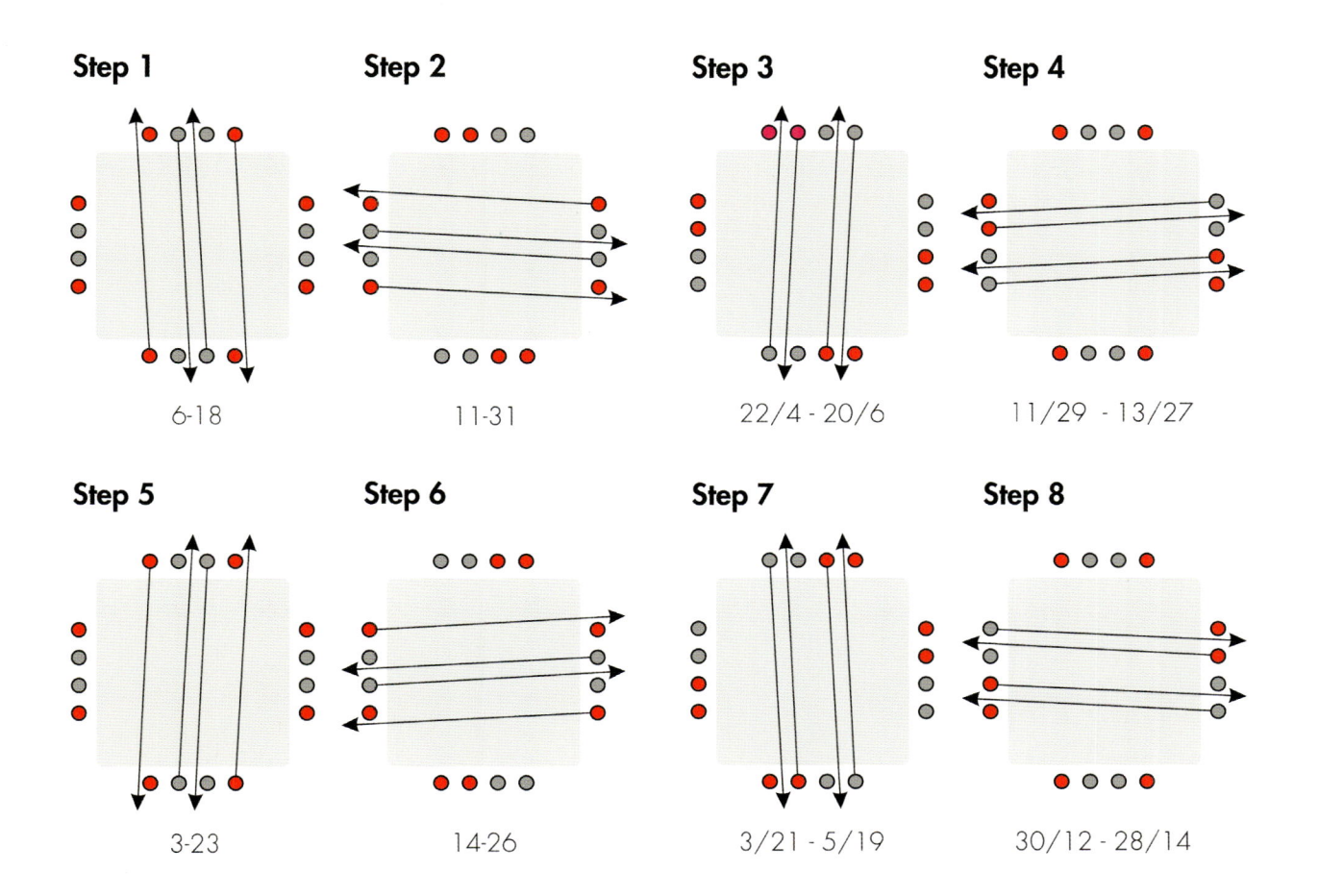

**Step 1**  6-18

**Step 2**  11-31

**Step 3**  22/4 - 20/6

**Step 4**  11/29 - 13/27

**Step 5**  3-23

**Step 6**  14-26

**Step 7**  3/21 - 5/19

**Step 8**  30/12 - 28/14

## Card Moves

| Clockwise | | Anticlockwise | |
|---|---|---|---|
| 6-18 | 14-26 | 3-23 | 11-31 |
| 5-6 | 13-14 | 4-3 | 12-11 |
| 20-5 | 28-13 | 21-4 | 29-12 |
| 4-20 | 12-28 | 5-21 | 13-29 |
| 3-4 | 11-12 | 6-5 | 14-13 |
| 22-3 | 30-11 | 19-6 | 27-14 |

On completing each column, reposition the threads to their set-up slots.

# 24.1
## 24-strand Chevron Braid

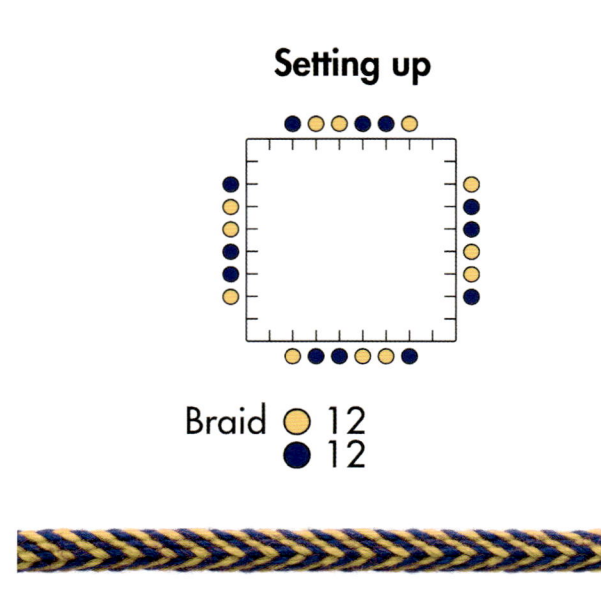

**Setting up**

Braid ○ 12
● 12

24.1.1a  Two-color chevron.

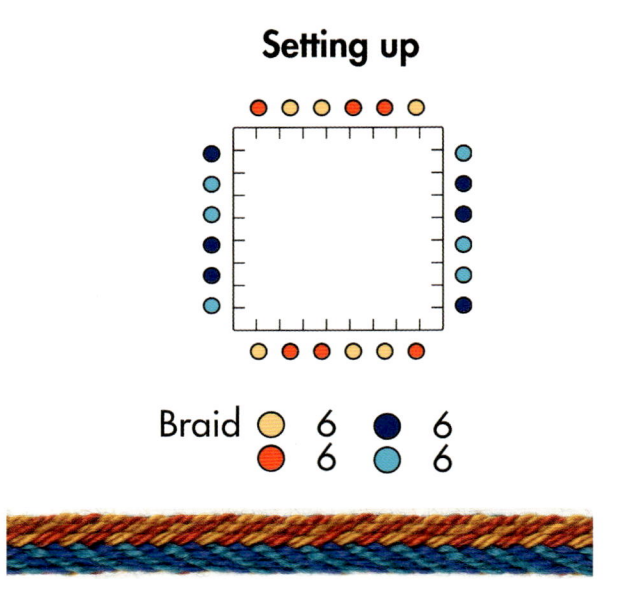

**Setting up**

Braid ○ 6  ● 6
● 6  ● 6

24.1.1c  Four-color chevron.

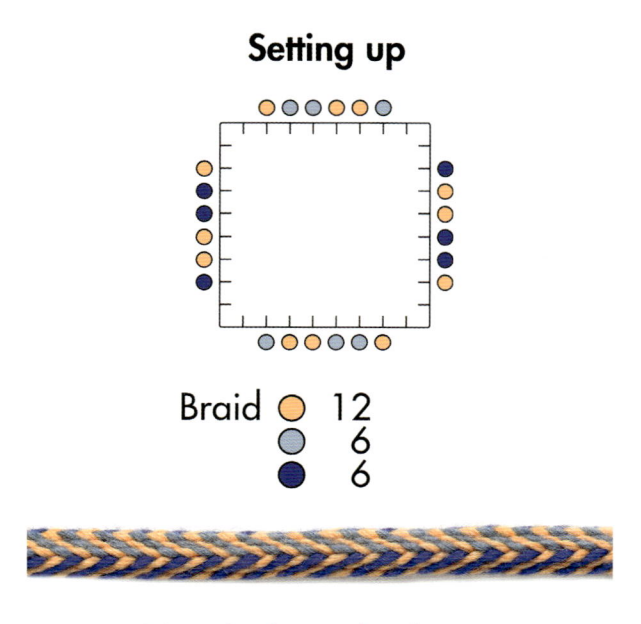

The chevron pattern is the basic structure on which 24-strand sling braids are based. Practice making this braid in order to memorize the hand movements for the Core Stand Method and the number sequences for the Card Method. For all designs shown, repeat Steps 1 and 2.

**Setting up**

Braid ○ 12
● 6
● 6

24.1.1b  Three-color chevron.

*Working Instructions for Design 24.1.*

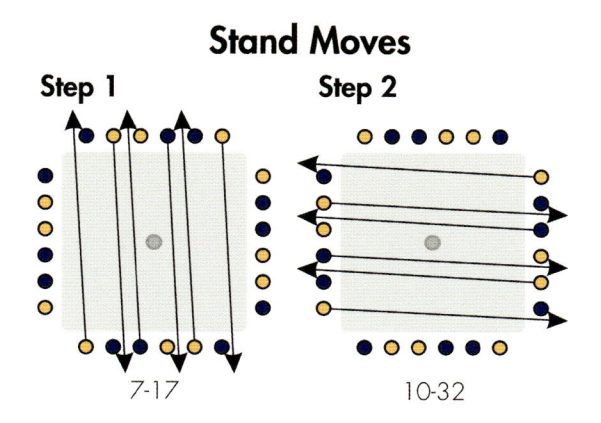

## Stand Moves

**Step 1**

7-17

**Step 2**

10-32

## Card Moves

| 7-17 | 10-32 |
|------|-------|
| 6-7  | 11-10 |
| 19-6 | 30-11 |
| 5-19 | 12-30 |
| 4-5  | 13-12 |
| 21-4 | 28-13 |
| 3-21 | 14-28 |
| 2-3  | 15-14 |
| 23-2 | 26-15 |

On completing each column, reposition the threads to their set up slots.

# 24.2
## Chevron Braid with Three Colors

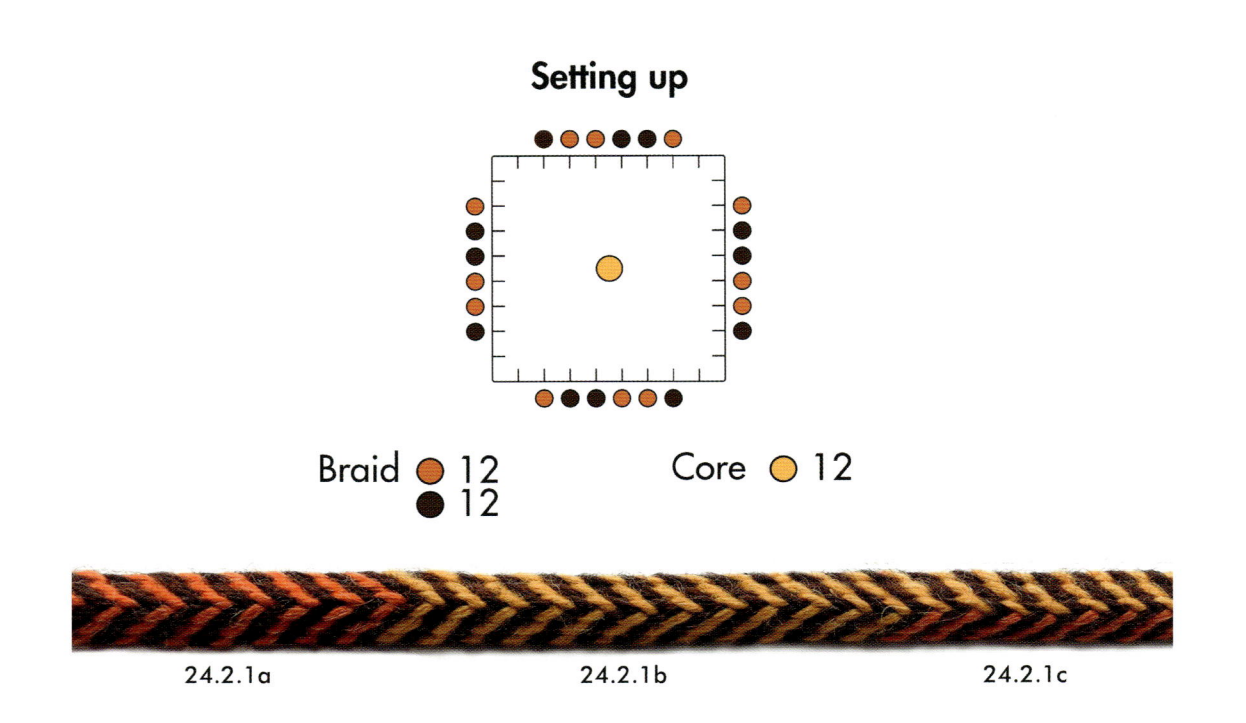

**Setting up**

Braid ● 12
● 12

Core ● 12

24.2.1a        24.2.1b        24.2.1c

### 24.2.1a Two-color chevron.
Place the gold threads on the harness and arrange the orange and brown threads around the braiding stand. Without changing the core, make a length of the orange and brown chevrons by repeating Steps 1 and 2.

### 24.2.1b Two-color chevrons.
To make the first core change, work Steps 1–6, swapping orange for gold. To make a length of brown and gold braid, repeat Steps 7 and 8.

### 24.2.1c Three-color chevron.
To create a brown braid with a bi-colored background of gold and orange, work Steps 1–6, but only change the N/S threads to orange. Make no core changes on the even numbered E/W steps.

### 24.2.1d Three-color chevron.
To change the braid to orange with a bi-colored background of gold and brown, change the E/W brown threads for orange.

### 24.2.1e Two-color chevron.
To make the remaining two-color combination, change the N/S brown threads for orange.

*The last of the six possible combinations, a yellow braid with a bi-colored background of orange and brown, is not shown.*

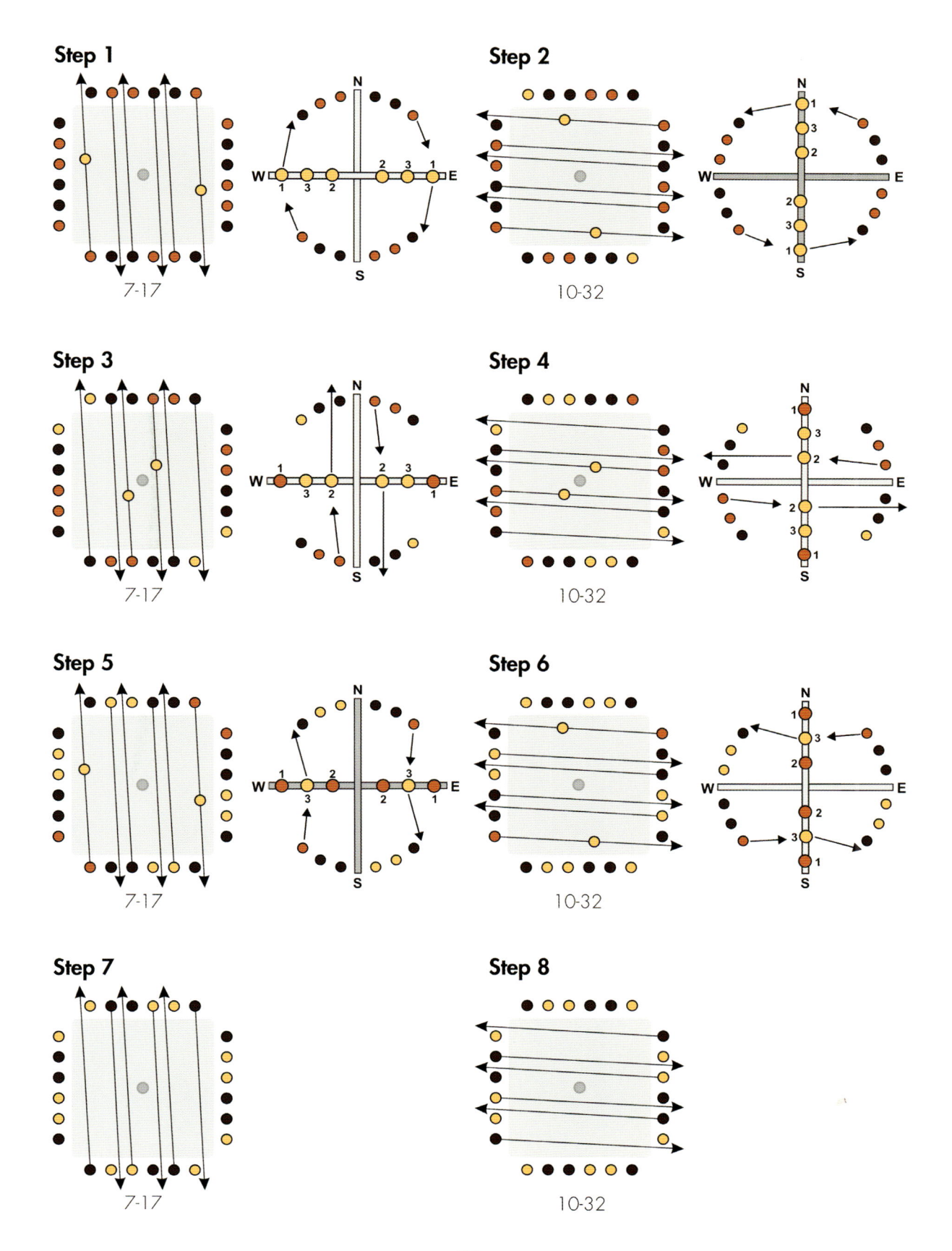

*Working Instructions for Design 24.2.1e*

**Step 1**

7-17

**Step 2**

10-32

**Step 3**

7-17

**Step 4**

10-32

**Step 5**

7-17

**Step 6**

10-32

**Step 7**

7-17

**Step 8**

10-32

97

# 24.3
## Chevron Braid with Vertical Stripe

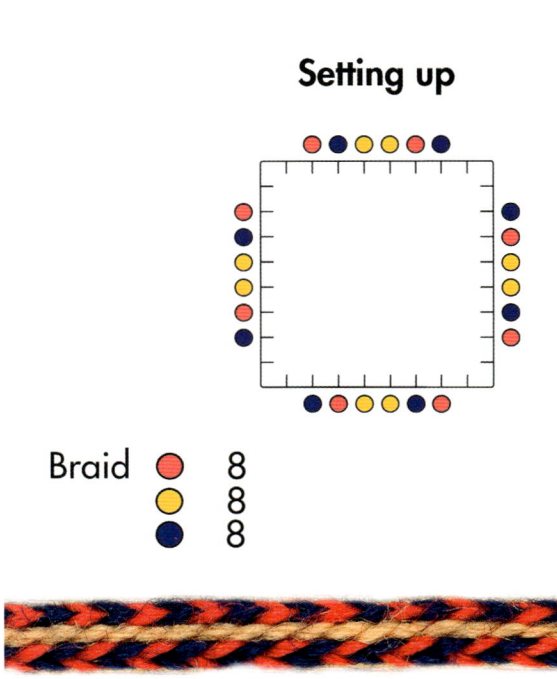

**Setting up**

Braid
- 🔴 8
- 🟡 8
- 🔵 8

**24.3.1** Three-color chevron with stripe. Repeat Steps 1-4.

**24.3.2** Three-color chevron.

This design was contributed by Laverne Waddington. She learned to make it in Peru where it is called "Palma braid." It uses the customary rotational movements of 24-strand chevron braids but uses an alternate color arrangement. It has no core. See Chapter 7 for details of how it is used in a sling.

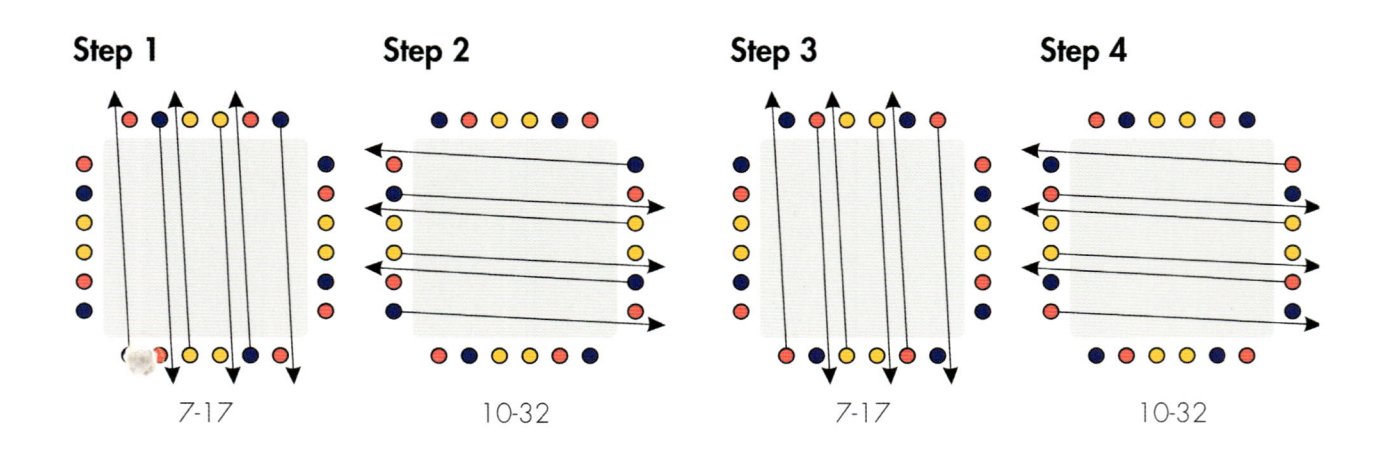

**Step 1** 7-17

**Step 2** 10-32

**Step 3** 7-17

**Step 4** 10-32

# 24.4
## Diamond Braid with Variations

The specific steps used to braid Design 24.4.1 are detailed in figures 7.2a–7.2d that follow. In order to help the braider develop a stronger understanding of the role that inversion and rotation moves play in creating pattern, the instructions below explore some of the possible ways in which these moves can be combined to create patterns.

**24.4.2 Two-color chevrons and arrows.**

Repeated inversion moves create textured stripes. Interesting pattern variations can be created when stripes are interspersed with A-shaped or V-shaped chevrons. To make this braid:

1. Repeat Steps 1 and 2 to make a length of chevrons.
2. To make a stripe, repeat Steps 3, 4, 7, and 8 for a further length, ending with Steps 3 and 4.
3. Repeat Steps 5 and 6 to close the lozenge and make a border of chevrons as seen in 24.4.4.
4. Continue experimenting with various combinations of steps and repetitions to create designs like 24.4.3–24.4.5.

**Setting up**

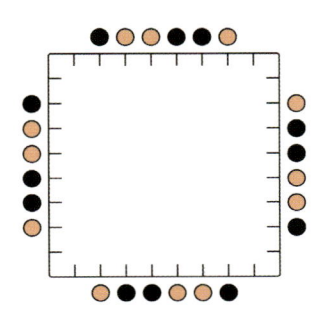

Braid ○ 12
● 12

**24.4.1 Two-color diamond with chevron borders.**

**Stripe and chevron combinations.**

24.3

24.4

24.5

1. To make a V-shaped chevron, repeat rotation Steps 1 and 2.
2. To introduce a diamond, work inversion Steps 3 and 4. Repeat rotation Steps 5 and 6 to complete the diamond and make a band of A-shaped chevrons.
3. To make another diamond pattern, work Steps 7 and 8 once, then repeat Steps 1 and 2 to complete the diamond and create another band of chevrons.

*Working Instructions for Designs 24.4.1.*

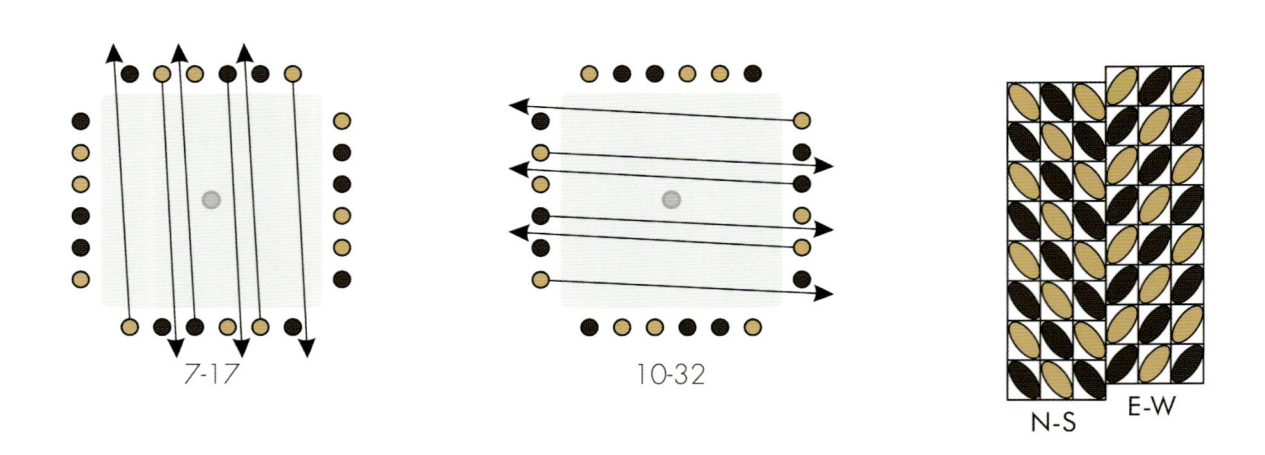

7-17

10-32

N-S  E-W

Figure 7.2a
Steps 1 and 2: Rotational moves are repeated to make a series of V-shaped chevrons.

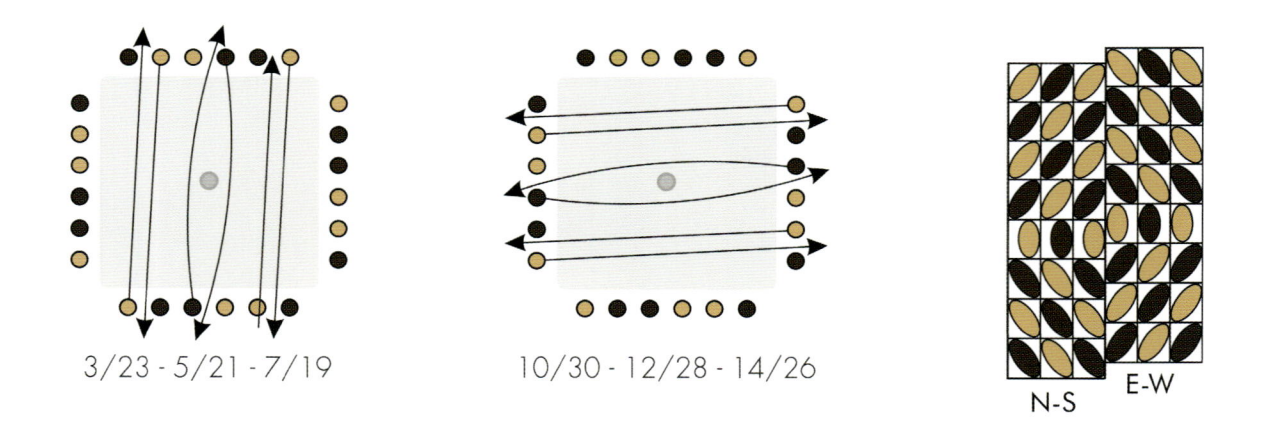

3/23 - 5/21 - 7/19

10/30 - 12/28 - 14/26

N-S  E-W

Figure 7.2b
Steps 3 and 4: N/S and E/W inversion steps
are worked once to invert the structure and make the eye of the diamond.
(The diamond pattern appears on the S/E corner of the braid.)

2-24

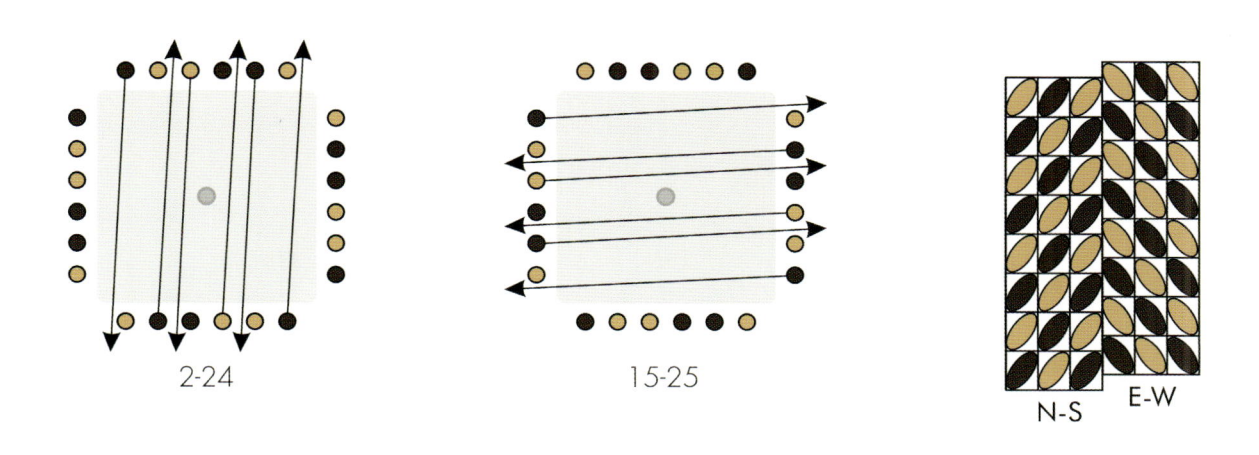

15-25

N-S  E-W

Figure 7.2c
Steps 5 and 6: The inversion steps have switched the position of the upper and lower threads.
These steps make the top half of the diamond pattern, an A-shaped chevron.

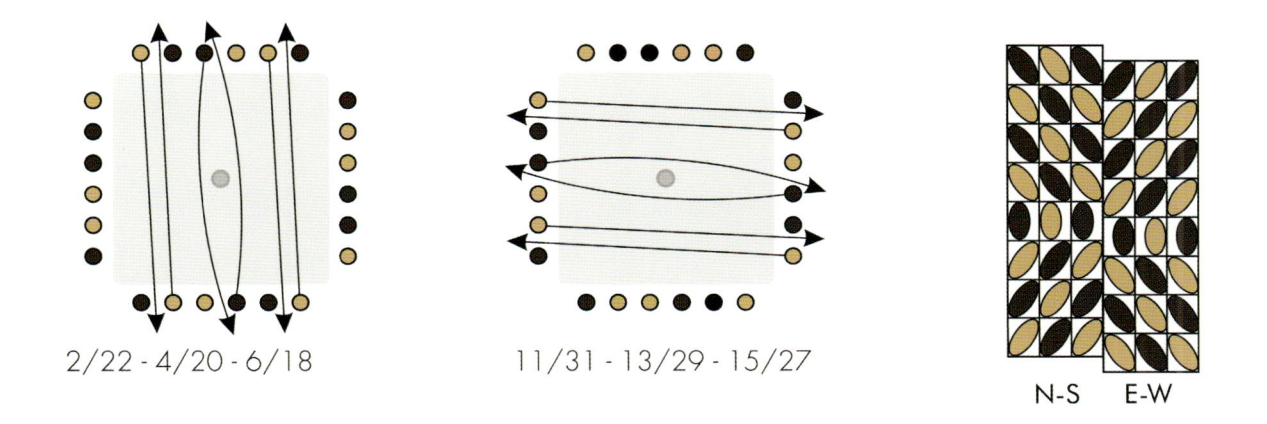

2/22 - 4/20 - 6/18

11/31 - 13/29 - 15/27

N-S  E-W

Figure 7.2e
Steps 7 and 8: The structure is inverted once again, making the eye of another diamond.
(Note that in Steps 7 and 8 the diamond appears on the N/W corner of the braid.)

# 24.5
## Chevron Braid with Elongated Pattern

★ Adjust braid before working step

**Setting up**

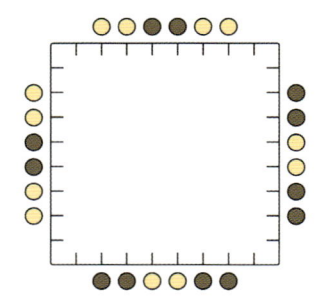

Braid ● 12
○ 12

**Setting up**

Braid ● 12   Core ● 12
○ 12

**24.5.1a** Two-color chevron with an elongated pattern.

This braid is included here because it is made with continuous inversion steps.
It does not require a core. Repeat Steps 1–8.

**24.5.1b** Two-color chevron.

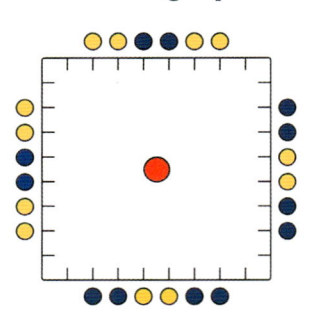

**24.5.2a** Three-color chevron made with core changes.

To make a blue and yellow braid, repeat Steps 1–8.

To change to a yellow and red braid, work Steps 9–16.
To continue this design, work Steps 1–8.

**24.5.2b** Three-color chevron made in medium contrast colors.

Color changes in this braid do not continue across a complete strip.
Therefore, colors that have a medium value contrast will soften the color change.

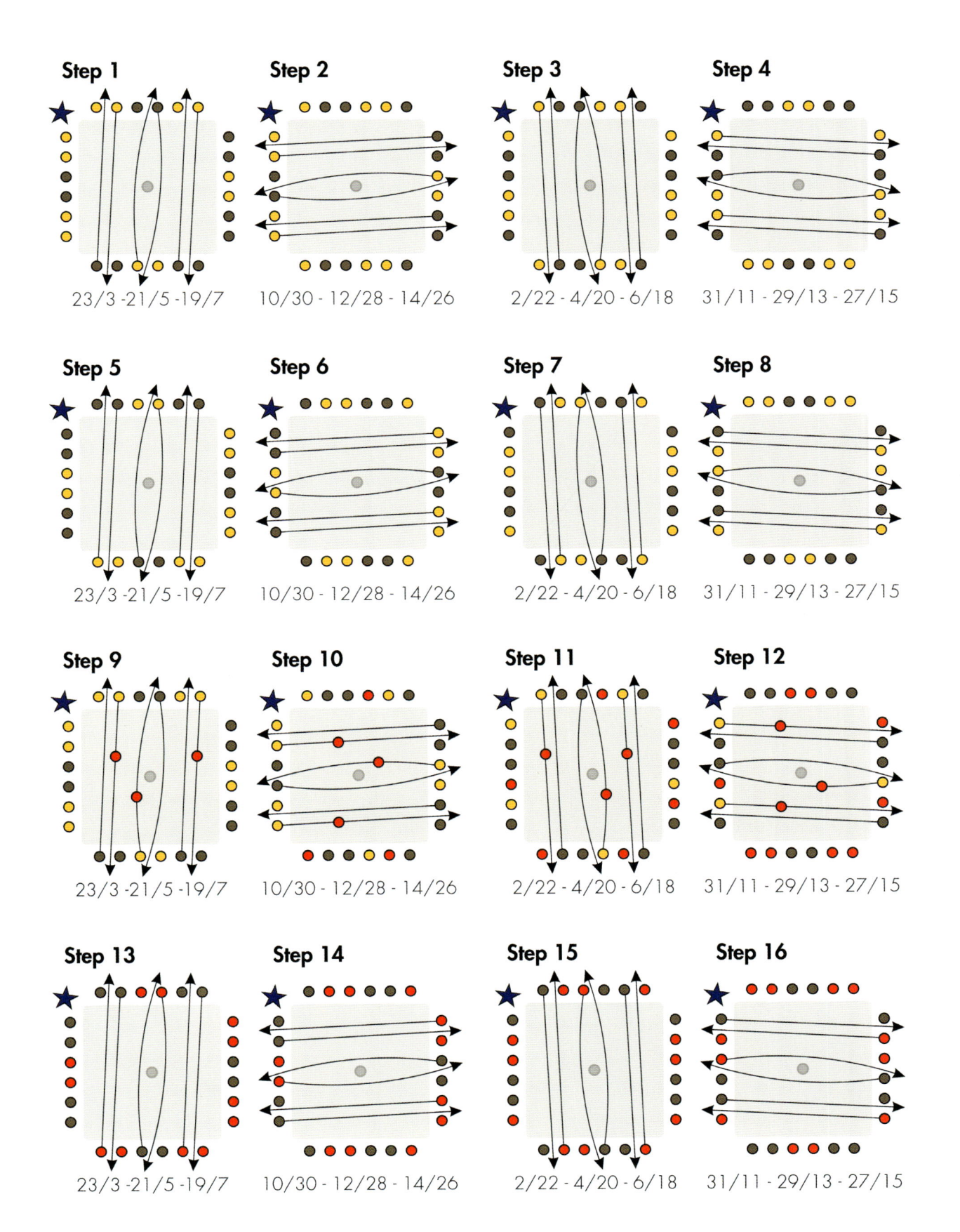

*Working Instructions for Designs 24.5.1 and 24.5.2.*

### Step 1
23/3 - 21/5 - 19/7

### Step 2
10/30 - 12/28 - 14/26

### Step 3
2/22 - 4/20 - 6/18

### Step 4
31/11 - 29/13 - 27/15

### Step 5
23/3 - 21/5 - 19/7

### Step 6
10/30 - 12/28 - 14/26

### Step 7
2/22 - 4/20 - 6/18

### Step 8
31/11 - 29/13 - 27/15

### Step 9
23/3 - 21/5 - 19/7

### Step 10
10/30 - 12/28 - 14/26

### Step 11
2/22 - 4/20 - 6/18

### Step 12
31/11 - 29/13 - 27/15

### Step 13
23/3 - 21/5 - 19/7

### Step 14
10/30 - 12/28 - 14/26

### Step 15
2/22 - 4/20 - 6/18

### Step 16
31/11 - 29/13 - 27/15

# 24.6
## Two-color Single and Continuous Diamonds

★ Adjust braid before working step

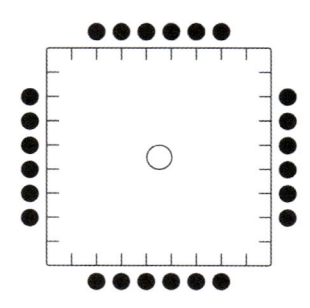

**Setting up**

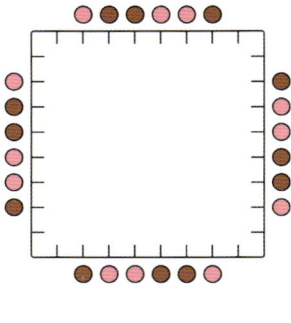

Braid ● 24   Core ○ 12

Braid ○ 12
       ● 12

### 24.6.1 Single diamonds.
Make a length of plain braid working Steps 1 and 2 without changing the core.
To make a single diamond pattern work Steps 1–16.

### 24.6.2 Continuous diamonds without a core.
Arrange the threads according to the setting up diagram above.
Repeat Steps 1–16 without changing the core.

### 24.6.3 Combined diamond patterns.
To make a braid with alternating single and continuous diamond patterns, first work Steps 1–16 of Design 24.6.1 changing the core, then work Steps 1–16 without changing the core to create a section of continuous diamonds. To finish the last diamond of the lozenge and return the braid to its background color, change the core in Steps 9–16.

*Working Instructions for Designs 24.6.1 to 24.6.3.*

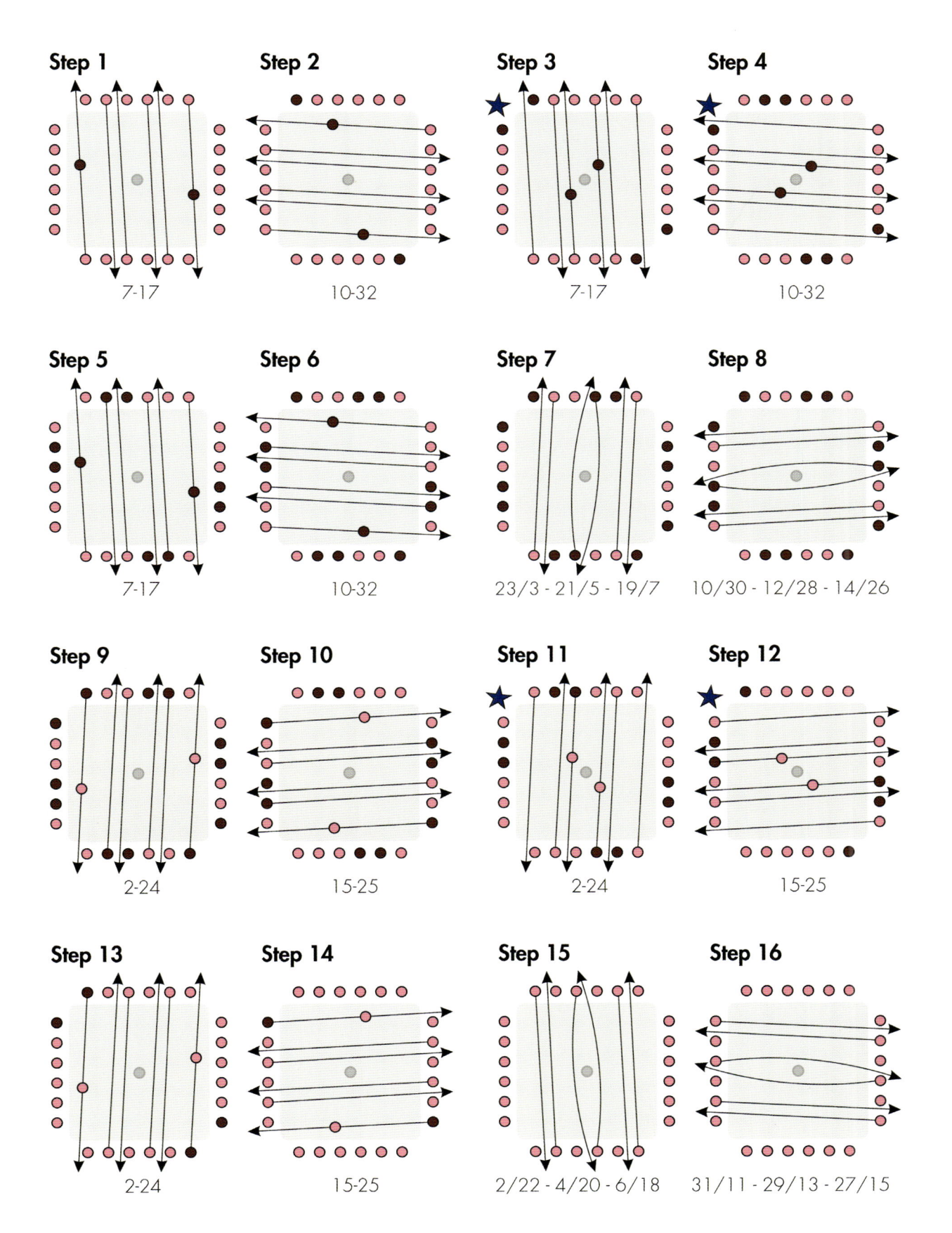

**Step 1** — 7-17
**Step 2** — 10-32
**Step 3** — 7-17
**Step 4** — 10-32
**Step 5** — 7-17
**Step 6** — 10-32
**Step 7** — 23/3 - 21/5 - 19/7
**Step 8** — 10/30 - 12/28 - 14/26
**Step 9** — 2-24
**Step 10** — 15-25
**Step 11** — 2-24
**Step 12** — 15-25
**Step 13** — 2-24
**Step 14** — 15-25
**Step 15** — 2/22 - 4/20 - 6/18
**Step 16** — 31/11 - 29/13 - 27/15

105

# 24.7
## Two-color Single and Continuous Diamonds

★ Adjust braid before working step

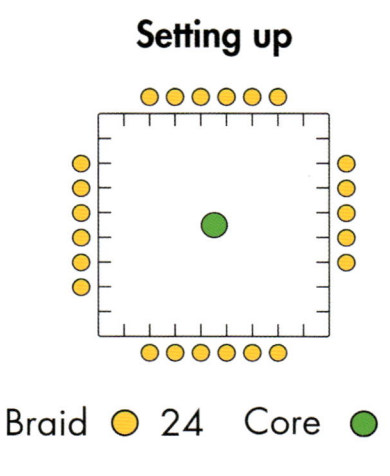

### Setting up

Braid ○ 24   Core ●

### Setting up

Braid ○ 12   ● 12

**24.7.1a Single diamonds.**

Repeat Steps 1 and 2 without changing the core to make a length of plain-colored braid.

To make the diamond pattern, work Steps 1–20.

To make a braid with alternating single and continuous diamonds patterns (not shown) first work Steps 1–20 with core changes for a single diamond, then work Steps 1–20 without core changes for a section of continuous diamonds.

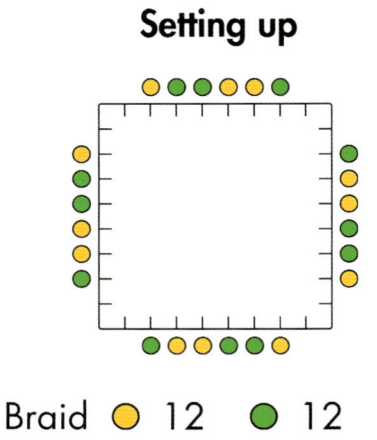

**24.7.2 Continuous diamonds without a core.**

Arrange the threads to the setting up diagram above. Repeat Steps 1–20.

**24.7.1b Single diamonds.**

**24.7.1c Side view.**

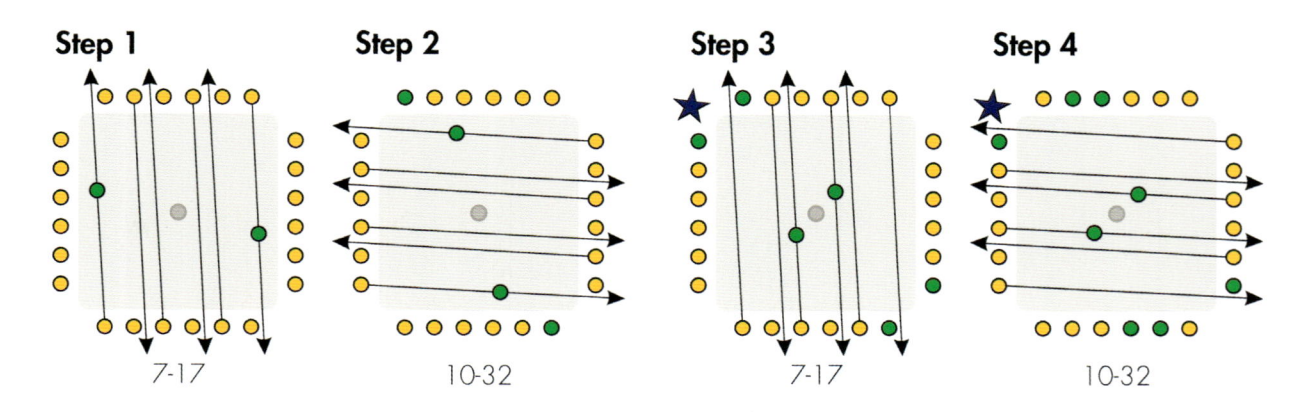

**Step 1**
7-17

**Step 2**
10-32

**Step 3**
7-17

**Step 4**
10-32

106

*Working Instructions for Designs 24.7.1 to 24.7.2.*

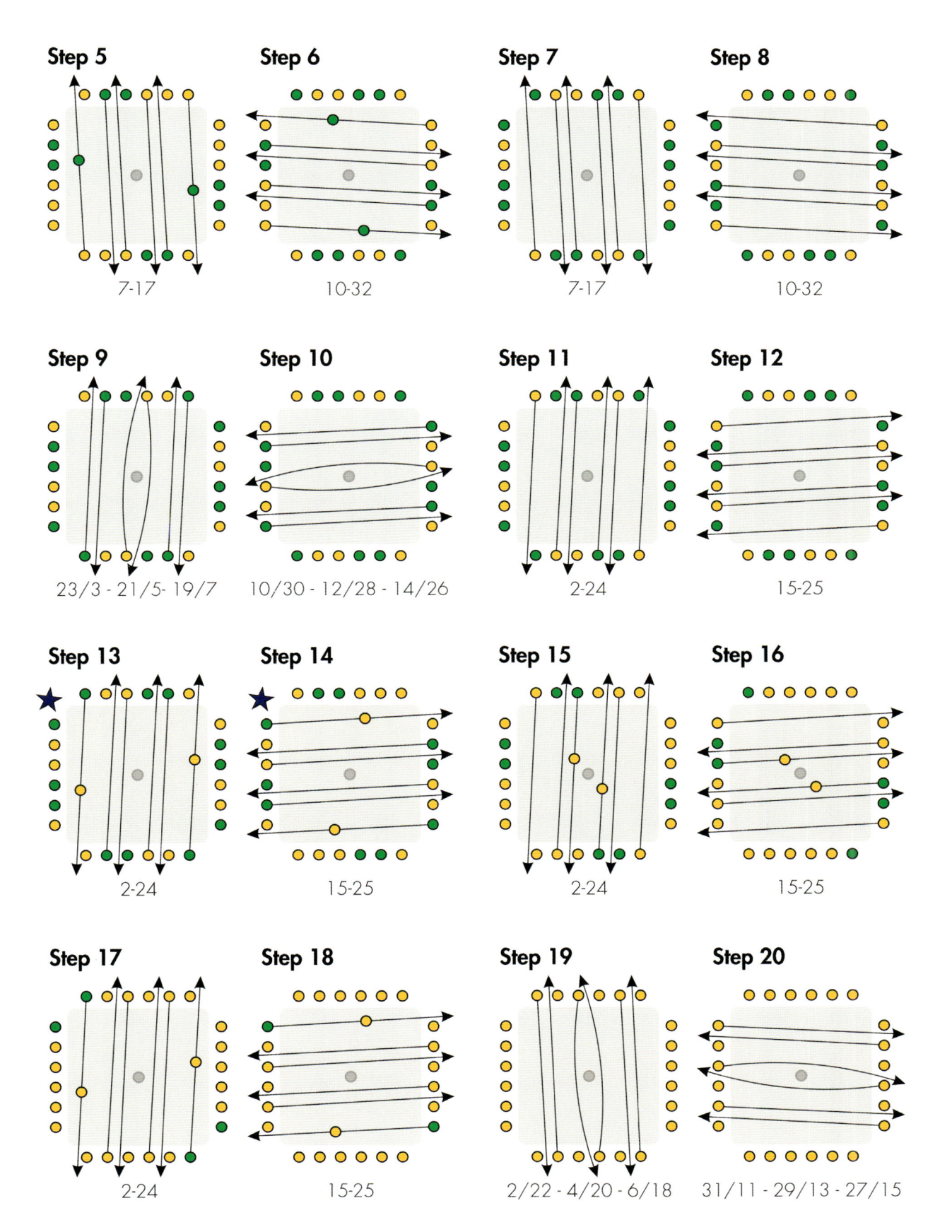

**Step 5**  7-17
**Step 6**  10-32
**Step 7**  7-17
**Step 8**  10-32

**Step 9**  23/3 - 21/5 - 19/7
**Step 10**  10/30 - 12/28 - 14/26
**Step 11**  2-24
**Step 12**  15-25

**Step 13**  2-24
**Step 14**  15-25
**Step 15**  2-24
**Step 16**  15-25

**Step 17**  2-24
**Step 18**  15-25
**Step 19**  2/22 - 4/20 - 6/18
**Step 20**  31/11 - 29/13 - 27/15

107

# 24.8
## Two-color Diamond Variations

★ Adjust braid before working step

**Setting up**

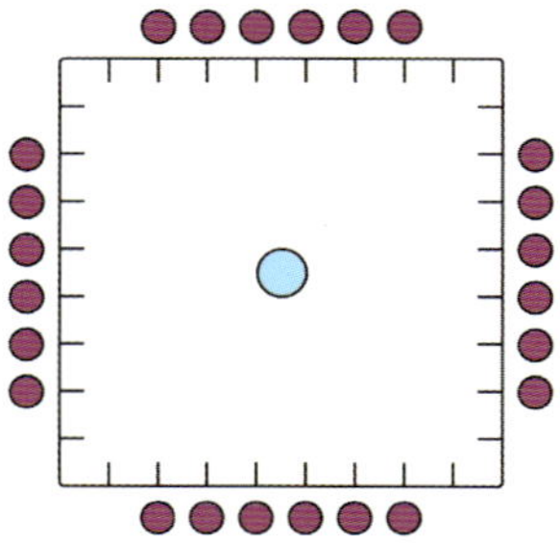

Braid ● 24   Core ○

### 24.8.1 Six diamond pattern.
Repeat Steps 1 and 2 without changing the core to make a length of plain-colored braid.

To make a six diamond pattern, work Steps 1–36.

**Setting up**

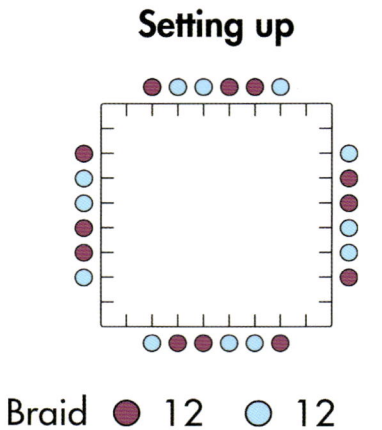

Braid ● 12   ○ 12

### 24.8.2 Continuous diamonds without a core.
Arrange the threads to the setting up diagram and repeat Steps 11–26.

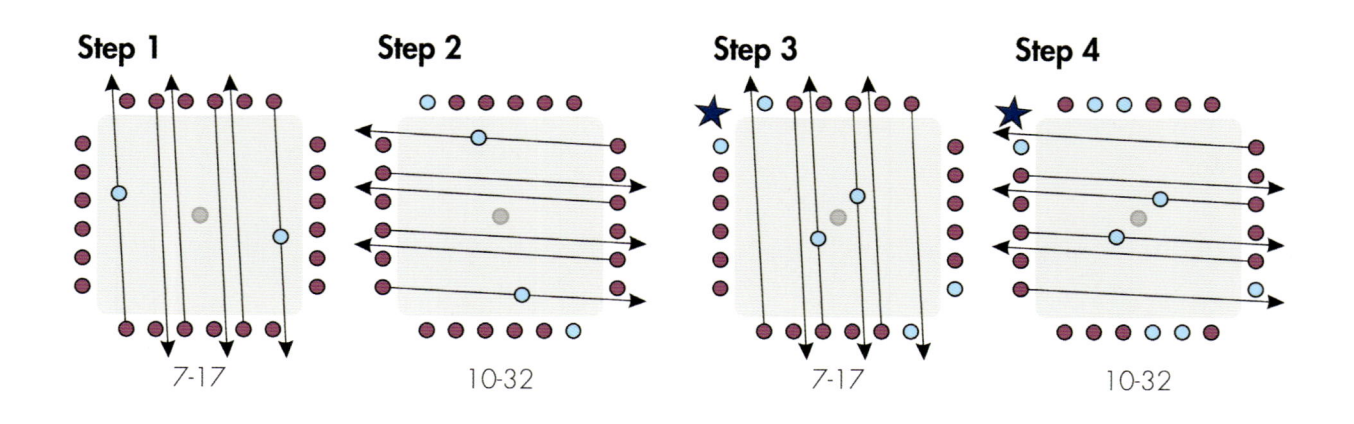

**Step 1**
7-17

**Step 2**
10-32

**Step 3**
7-17

**Step 4**
10-32

108

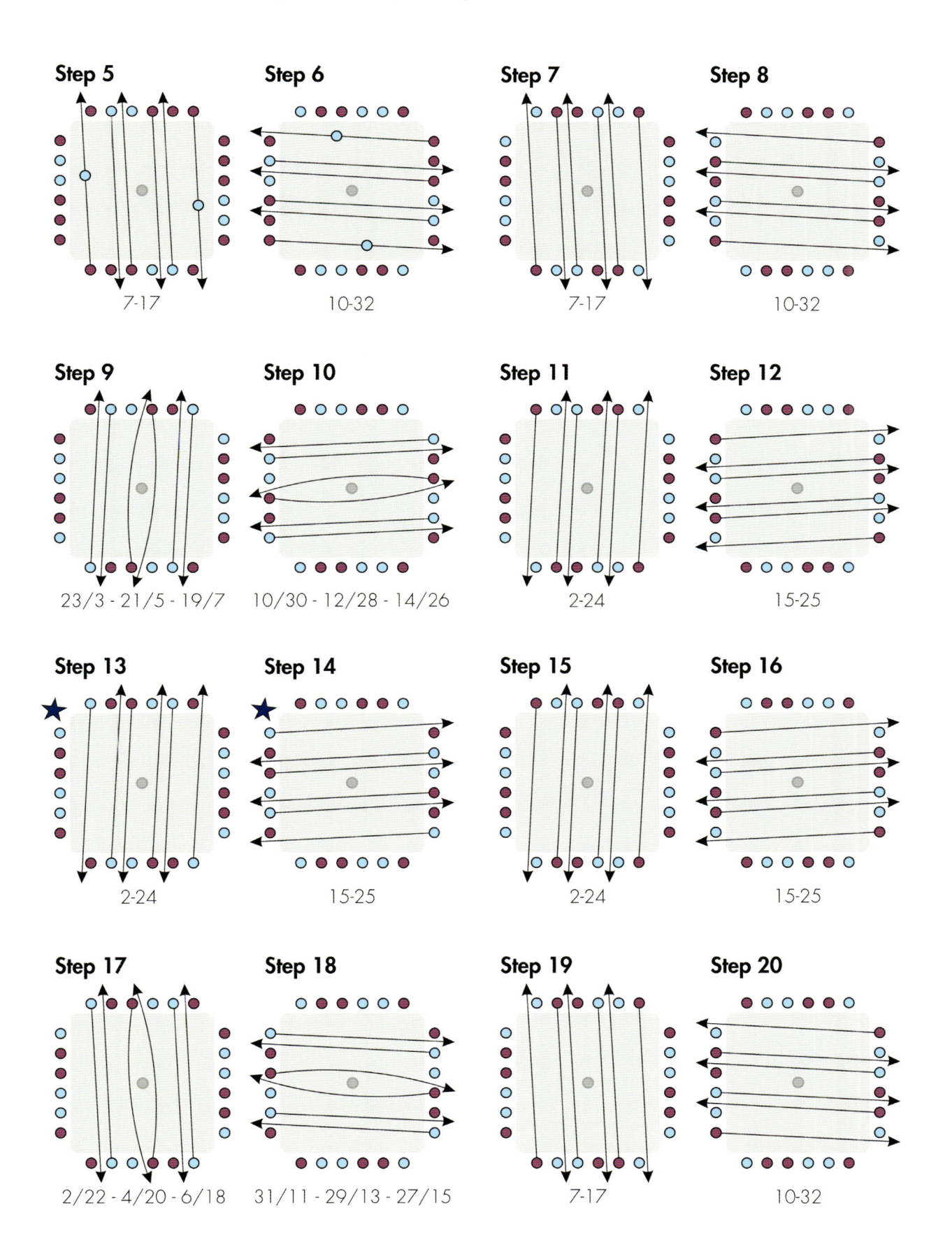

*Working Instructions for Designs 24.8.1 and 24.8.2 continued.*

**Step 5**
7-17

**Step 6**
10-32

**Step 7**
7-17

**Step 8**
10-32

**Step 9**
23/3 - 21/5 - 19/7

**Step 10**
10/30 - 12/28 - 14/26

**Step 11**
2-24

**Step 12**
15-25

**Step 13**
2-24

**Step 14**
15-25

**Step 15**
2-24

**Step 16**
15-25

**Step 17**
2/22 - 4/20 - 6/18

**Step 18**
31/11 - 29/13 - 27/15

**Step 19**
7-17

**Step 20**
10-32

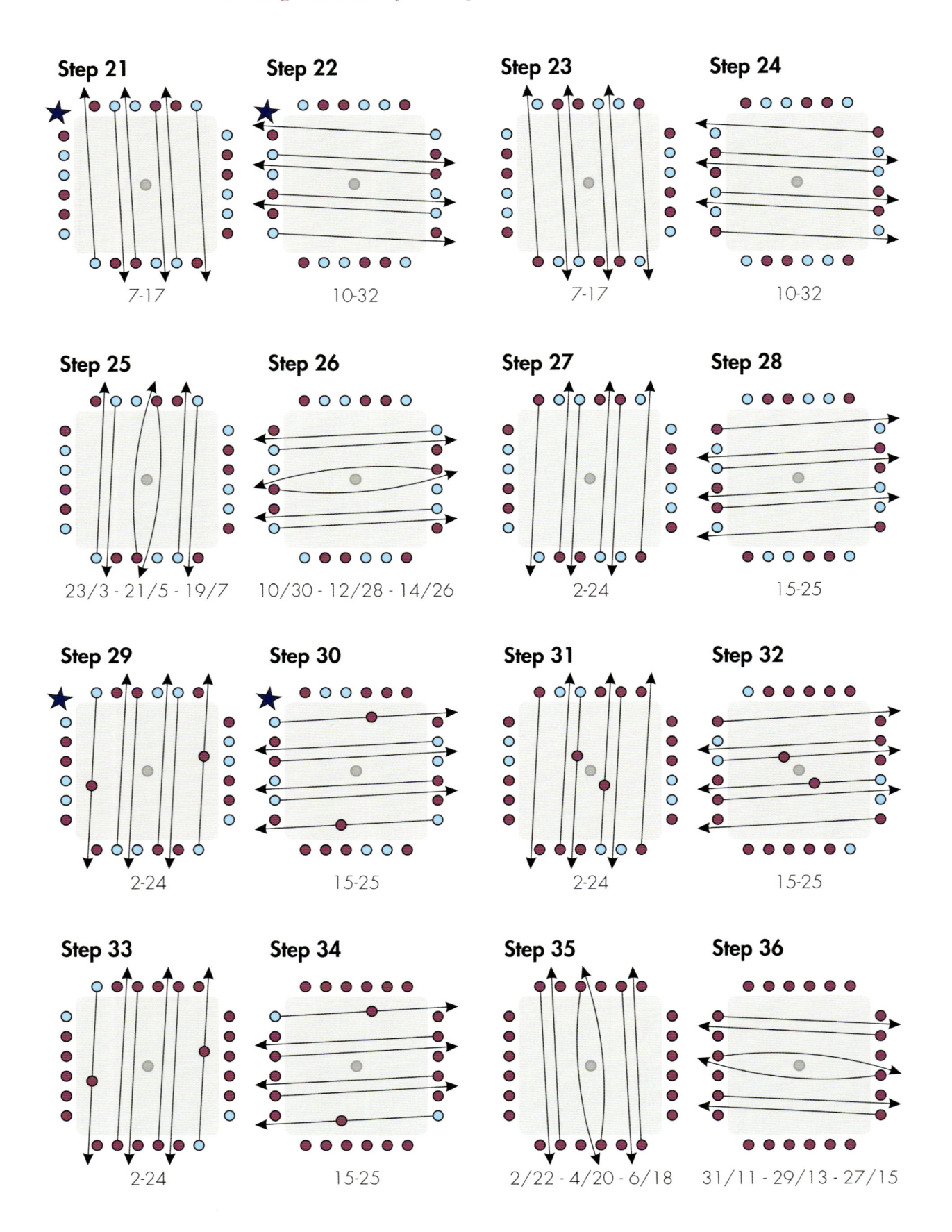

*Working Instructions for Designs 24.8.1 and 24.8.2 continued.*

**Step 21**
7-17

**Step 22**
10-32

**Step 23**
7-17

**Step 24**
10-32

**Step 25**
23/3 - 21/5 - 19/7

**Step 26**
10/30 - 12/28 - 14/26

**Step 27**
2-24

**Step 28**
15-25

**Step 29**
2-24

**Step 30**
15-25

**Step 31**
2-24

**Step 32**
15-25

**Step 33**
2-24

**Step 34**
15-25

**Step 35**
2/22 - 4/20 - 6/18

**Step 36**
31/11 - 29/13 - 27/15

# 24.9
## Two-color Linked and Continuous Diamonds

★ Adjust braid before working step

**Setting up**

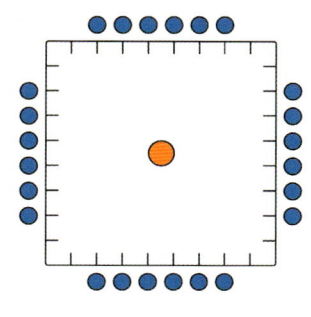

Braid ● 24   Core ○ 1

### 24.9.1 Linked diamonds.

Repeat Steps 1 and 2 without changing the core to make a length of plain-colored braid.

To make the linked diamond pattern, work Steps 1–32.

To make a braid with alternating linked and continuous diamonds, first work Steps 1–32, then repeat Steps 5–16.

**Setting up**

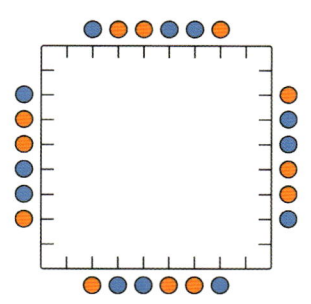

Braid ○ 12
○ 12

### 24.9.2 Continuous diamonds without a core.

Arrange the threads according to the setting up diagram above.
Repeat Steps 5–16.

111

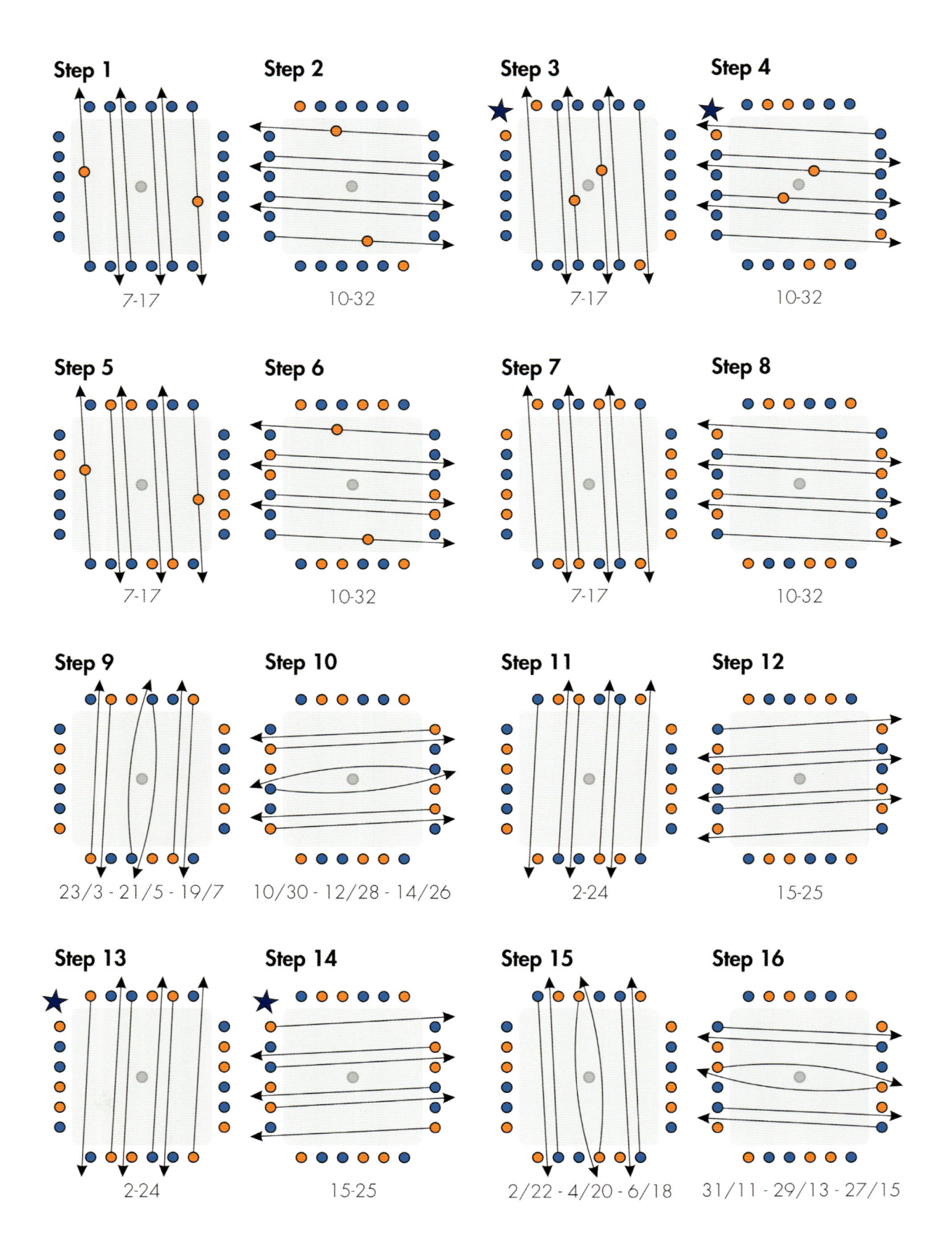

*Working Instructions for Designs 24.9.1 and 24.9.2.*

**Step 1**  7-17

**Step 2**  10-32

**Step 3**  7-17

**Step 4**  10-32

**Step 5**  7-17

**Step 6**  10-32

**Step 7**  7-17

**Step 8**  10-32

**Step 9**  23/3 - 21/5 - 19/7

**Step 10**  10/30 - 12/28 - 14/26

**Step 11**  2-24

**Step 12**  15-25

**Step 13**  2-24

**Step 14**  15-25

**Step 15**  2/22 - 4/20 - 6/18

**Step 16**  31/11 - 29/13 - 27/15

112

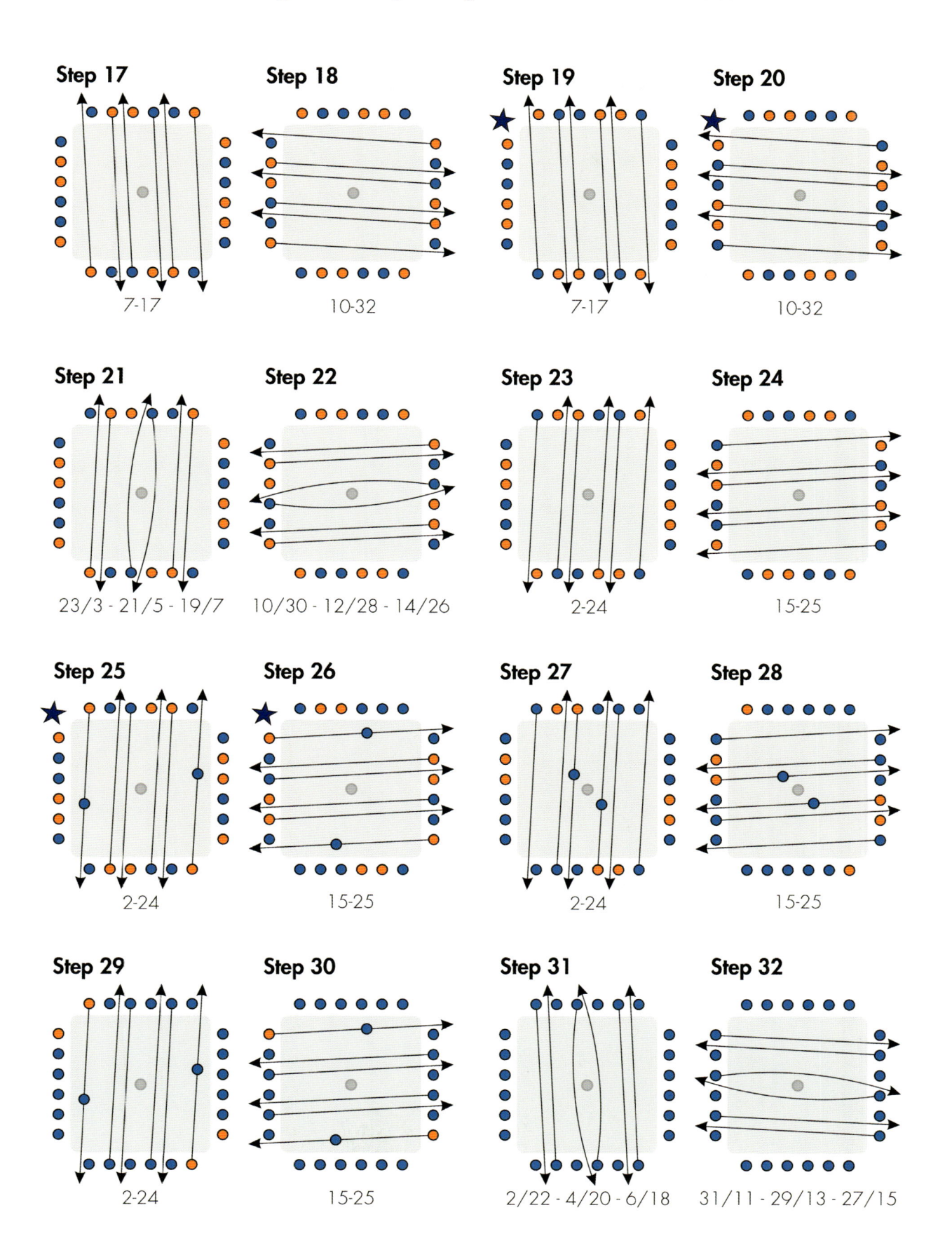

*Working Instructions for Designs 24.9.1 and 24.9.2 continued.*

**Step 17**
7-17

**Step 18**
10-32

**Step 19**
7-17

**Step 20**
10-32

**Step 21**
23/3 - 21/5 - 19/7

**Step 22**
10/30 - 12/28 - 14/26

**Step 23**
2-24

**Step 24**
15-25

**Step 25**
2-24

**Step 26**
15-25

**Step 27**
2-24

**Step 28**
15-25

**Step 29**
2-24

**Step 30**
15-25

**Step 31**
2/22 - 4/20 - 6/18

**Step 32**
31/11 - 29/13 - 27/15

# 24.10
## Two-color Double Diamonds with Variations

★ Adjust braid before working step

### Setting up

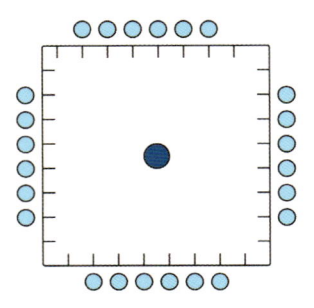

Braid ○ 24   Core ● 1

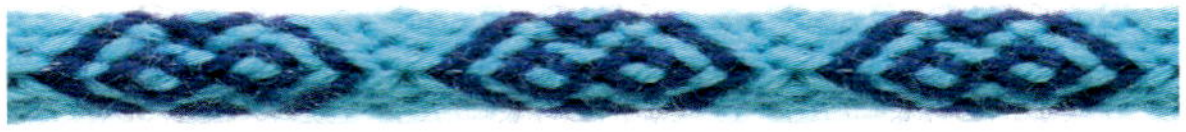

**24.10.1a** Side view showing W's and M's pattern.

As in other diamond designs, secondary patterns occur where two diamonds meet.

**24.10.1  Double Diamonds.**

Make a plain length of braid by repeating Steps 1 and 2 without changing the core.
To make the double diamond pattern, work Steps 1–28.

### Setting up

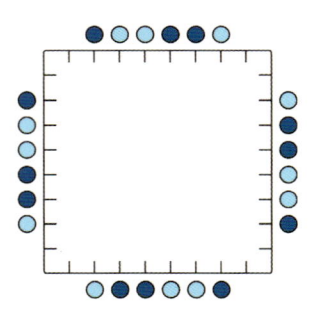

Braid ● 12   ○ 12

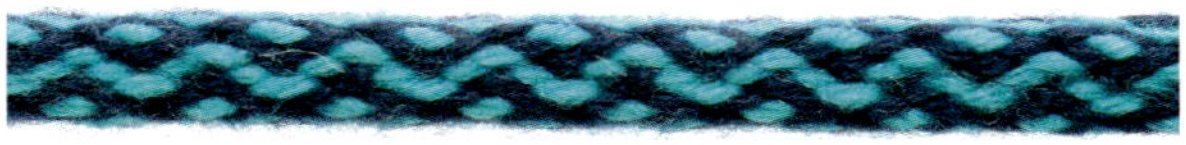

### Setting up

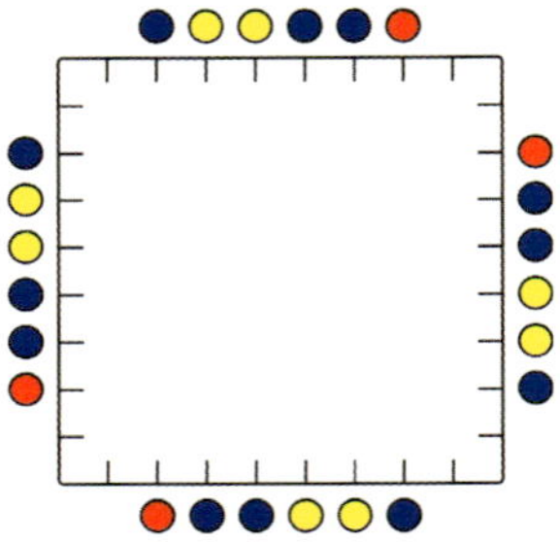

Braid ● 12
      ● 8
      ● 4

**24.10.3  Three-color small diamonds.**
Repeat Steps 7–14.

**24.10.2  Wave and spot pattern made without a core.**

Repeat Steps 7–14. This pattern can also be alternated with Designs 24.10.1 or 24.10.2 if a core is added.

114

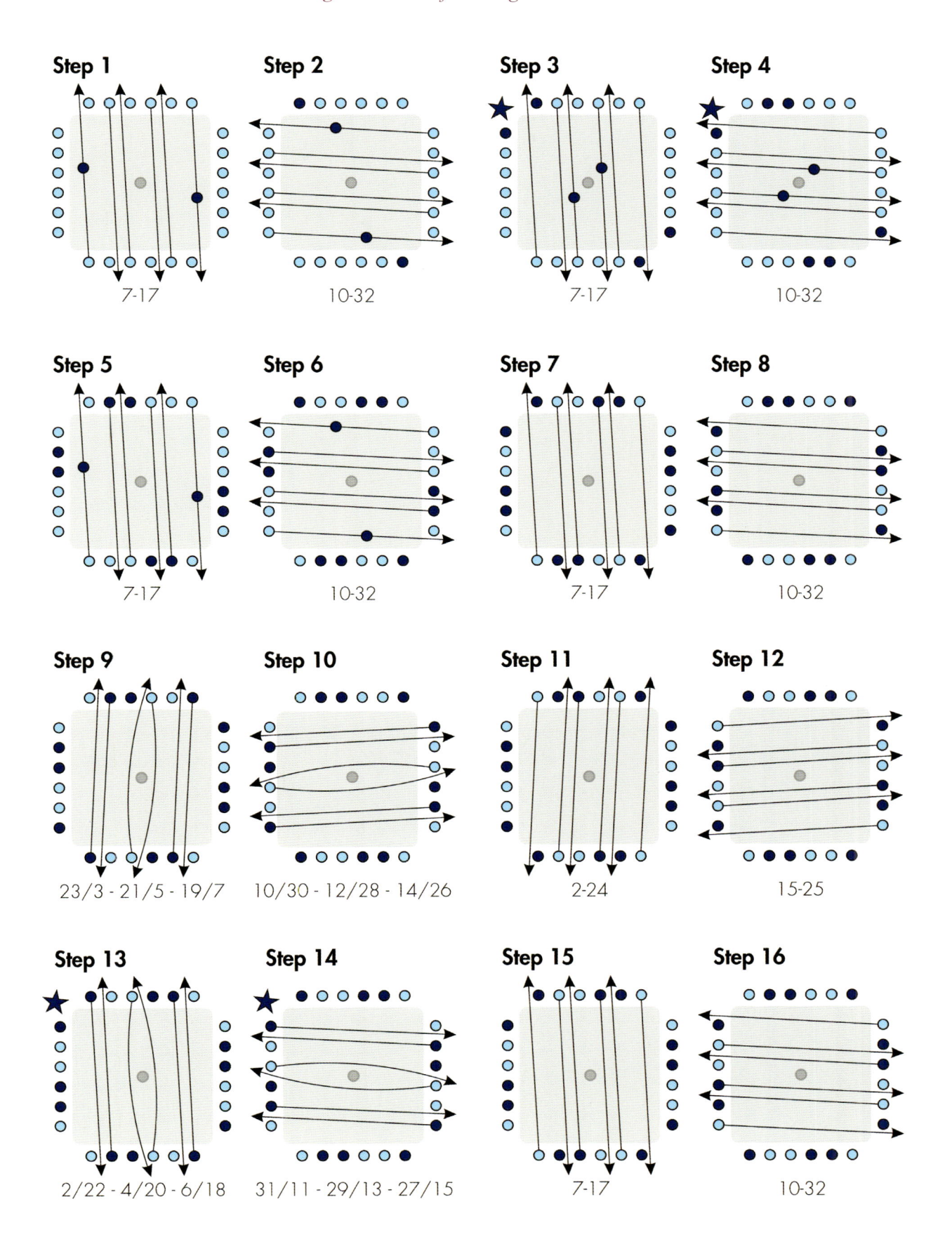

*Working Instructions for Designs 24.10.1–24.10.3.*

**Step 1**  7-17
**Step 2**  10-32
**Step 3**  7-17
**Step 4**  10-32
**Step 5**  7-17
**Step 6**  10-32
**Step 7**  7-17
**Step 8**  10-32
**Step 9**  23/3 - 21/5 - 19/7
**Step 10**  10/30 - 12/28 - 14/26
**Step 11**  2-24
**Step 12**  15-25
**Step 13**  2/22 - 4/20 - 6/18
**Step 14**  31/11 - 29/13 - 27/15
**Step 15**  7-17
**Step 16**  10-32

115

*Working Instructions for Designs 24.10.1–24.10.3 continued.*

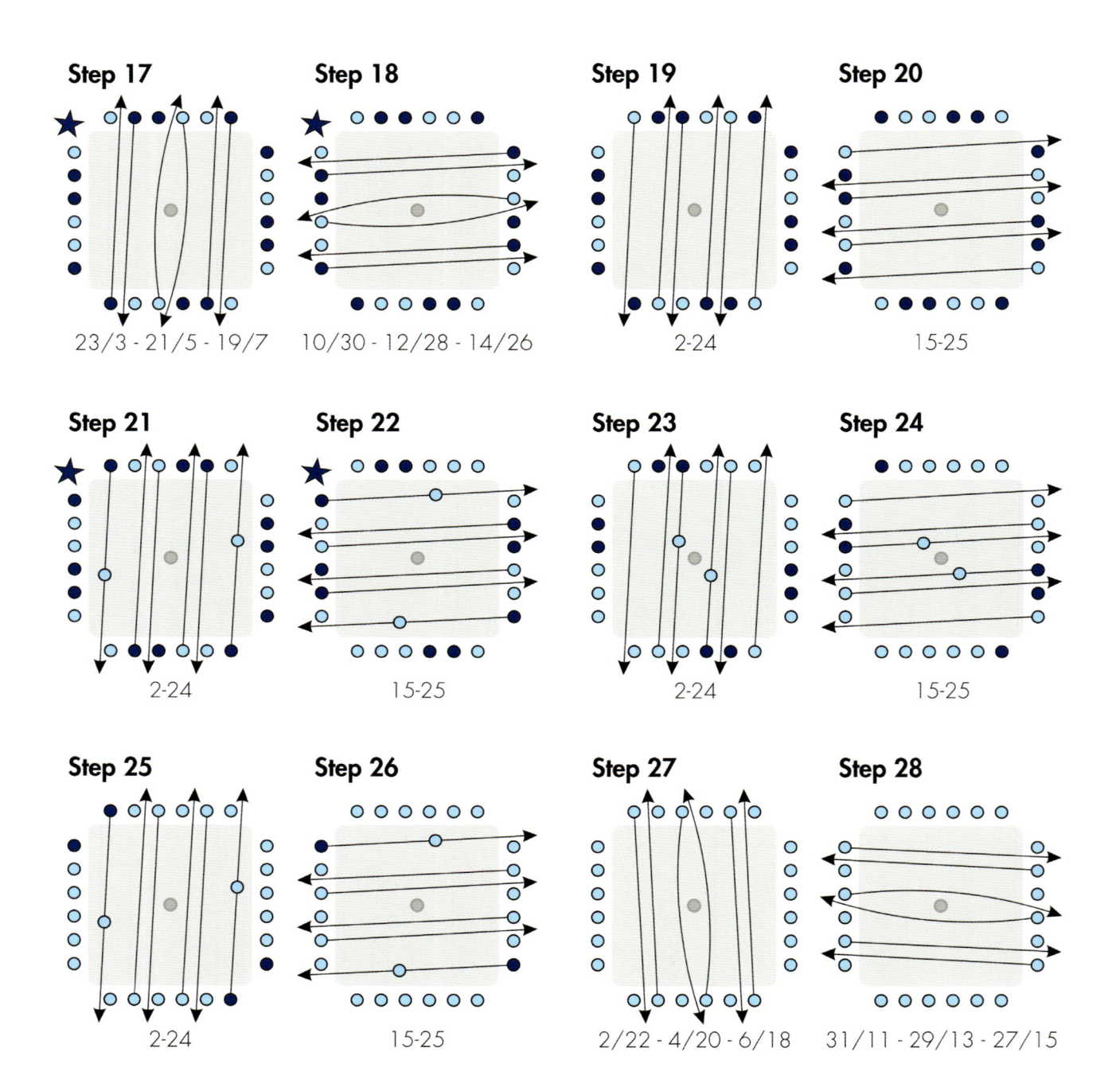

**Step 17**
23/3 - 21/5 - 19/7

**Step 18**
10/30 - 12/28 - 14/26

**Step 19**
2-24

**Step 20**
15-25

**Step 21**
2-24

**Step 22**
15-25

**Step 23**
2-24

**Step 24**
15-25

**Step 25**
2-24

**Step 26**
15-25

**Step 27**
2/22 - 4/20 - 6/18

**Step 28**
31/11 - 29/13 - 27/15

# 24.11
## Large Open Diamonds

★ Adjust braid before working step

**Setting up**

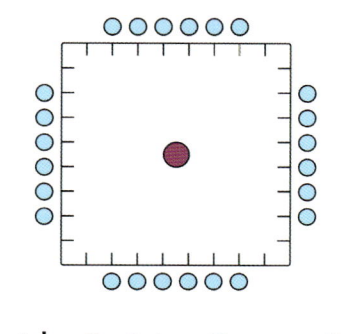

Braid ◯ 24   Core ● 12

**24.11.1a** Large open diamonds.
Thread exchanges at Steps 10 and 11 give this diamond its elongated shape.

Without changing the core, make a length of plain braid working Steps 1 and 2. To make the diamond design, work Steps 1–18.

**24.11.1b** Light pattern on a dark ground.

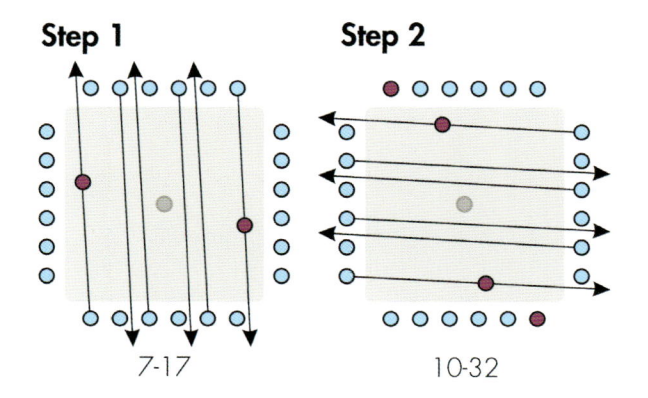

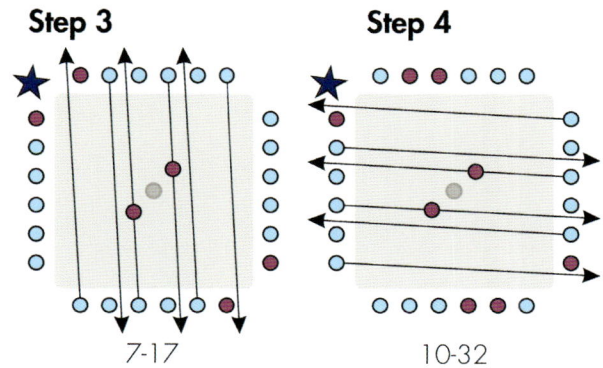

**Step 1**  7-17

**Step 2**  10-32

**Step 3**  7-17

**Step 4**  10-32

117

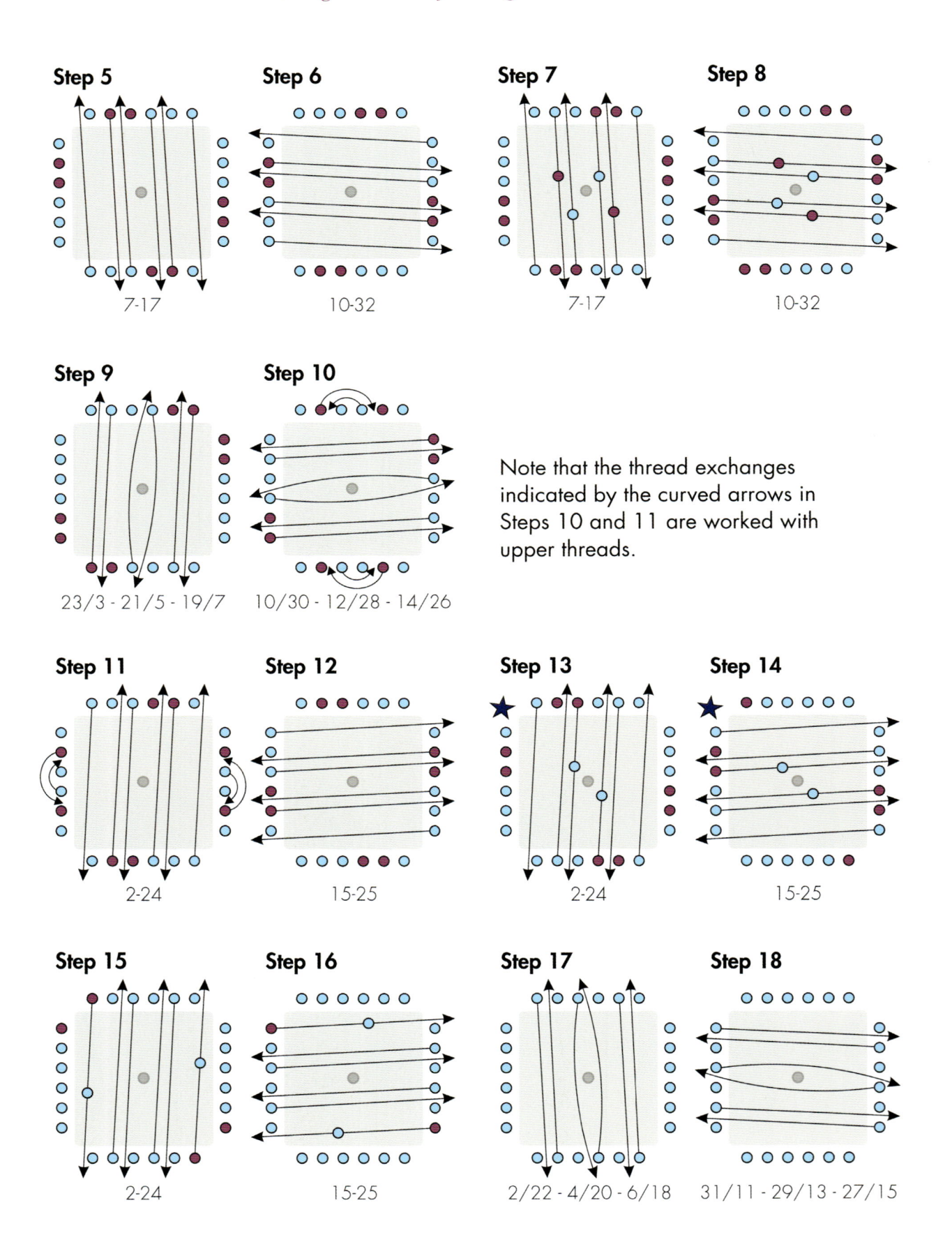

*Working Instructions for Design 24.11.1 continued.*

**Step 5**

7-17

**Step 6**

10-32

**Step 7**

7-17

**Step 8**

10-32

**Step 9**

23/3 - 21/5 - 19/7

**Step 10**

10/30 - 12/28 - 14/26

Note that the thread exchanges indicated by the curved arrows in Steps 10 and 11 are worked with upper threads.

**Step 11**

2-24

**Step 12**

15-25

**Step 13**

2-24

**Step 14**

15-25

**Step 15**

2-24

**Step 16**

15-25

**Step 17**

2/22 - 4/20 - 6/18

**Step 18**

31/11 - 29/13 - 27/15

118

# 24.12
## Diamonds with Three Colors Pattern

★ Adjust braid before working step

**Setting up**

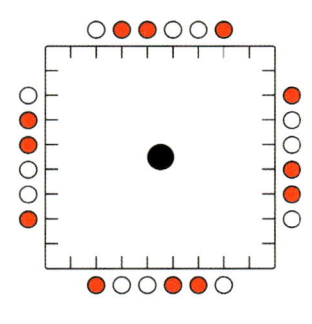

Braid ● 12   Core ● 12
○ 12

24.12.1 R&W "V" chevrons

24.12.2 Alternating R&W and B&W diamonds.

24.12.3 R&W diamonds.

24.12.4 R&W diamonds with black eyes

24.12.5 Small R&W diamonds with black eye.

24.12.6 R&W "A" chevrons.

24.12.7 B&W diamonds

24.12.8 Three-color diamonds.

The instructions for Design 24.12 are presented in two ways. The 16 steps used to make Design 24.12.2 are detailed on the next page. To make the other designs, refer to Table 5.1 that follows.

The table shows how to recombine the steps for Design 24.12.2 to create the other seven patterns. It also tells how to transition from one design to another. Not all possible combinations are shown but you can use the information in the table to deduce how to create additional designs.

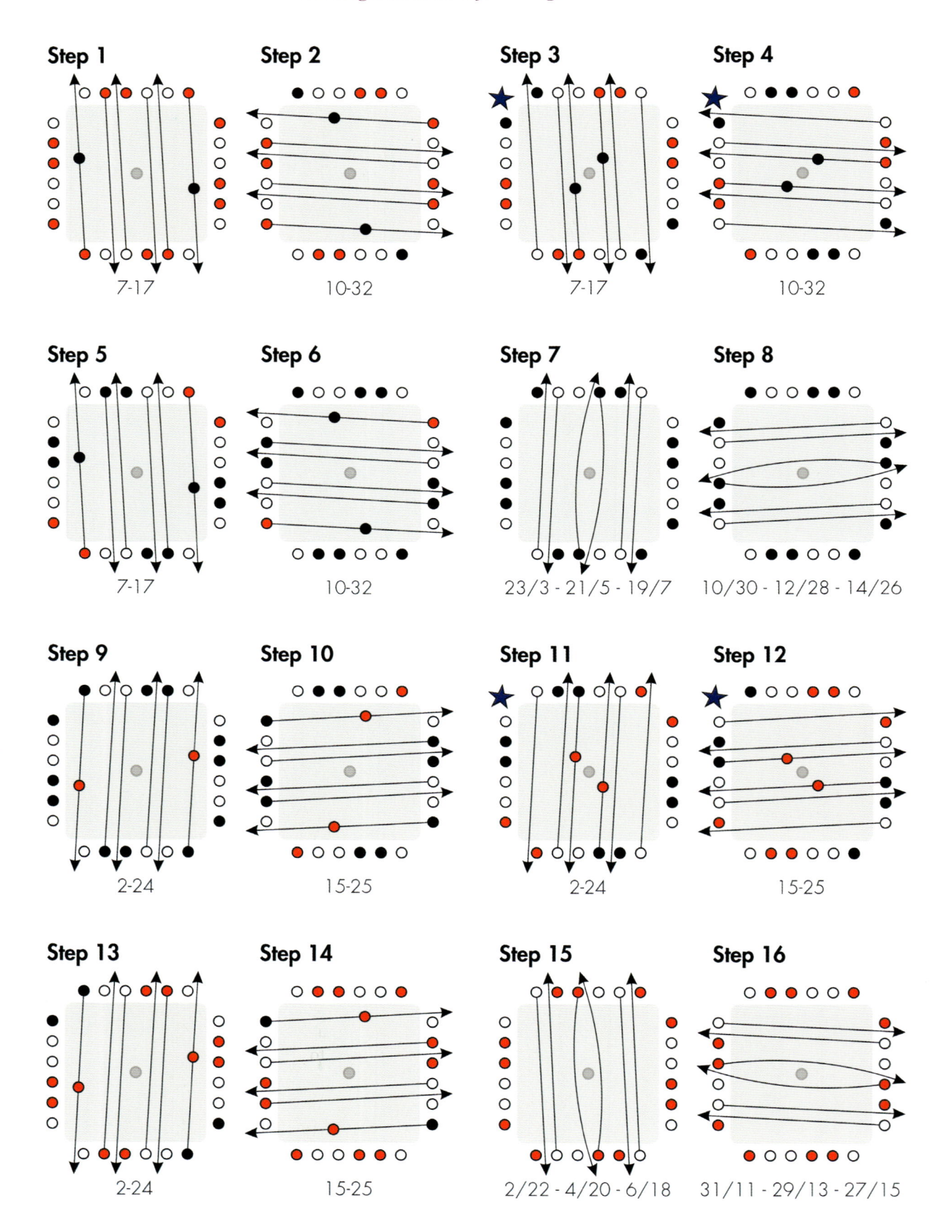

*Working Instructions for Designs 24.12.1.*

Table 5.1

| | Description | Instructions | 1-2 | 3-4 | 5-6 | 7-8 | 9-10 | 11-12 | 13-14 | 15-16 | |
|---|---|---|---|---|---|---|---|---|---|---|---|
| 24.12.1 | R&W "V" Chevrons | Repeat Steps 1&2 without changing the core | NCC | — | — | — | — | — | — | — | End chevron with light color in the SW and NE corners |
| 24.12.2 | Alternating R&W and B&W diamonds | Repeat Steps 1-16 | CC | CC | CC | INV | CC | CC | CC | INV | End sequence with Step 16 |
| 24.12.3 | R&W diamonds | Repeat Steps 1-16 | NCC | NCC | NCC | INV | NCC | NCC | NCC | INV | End sequence with Step 16 |
| 24.12.4 | R&W diamonds with Black eye | Work row 1, then repeat row 2 | 1. NCC<br>2. CC | NCC<br>CC | CC<br>CC | INV<br>INV | CC<br>CC | NCC<br>CC | CC<br>CC | INV<br>INV | End sequence with Step 16 |
| 24.12.5 | Small diamonds | Repeat Steps 1&2, 7&8, 9&10, 15&16 | NCC | — | — | INV | NCC | — | — | INV | End sequence with Step 16 |
| 24.12.6 | R&W "A" chevrons | Work row 1, then repeat row 2 | 1. CC<br>2. — | —<br>— | —<br>— | INV<br>— | —<br>NCC | —<br>— | —<br>— | —<br>— | End sequence with Step 10 |
| 24.12.7 | Convert to B&W diamonds | Work Steps 9 to 16<br>Repeat Steps 1 to 16 | 1. —<br>2. NCC | —<br>NCC | —<br>NCC | —<br>INV | CC<br>NCC | CC<br>NCC | CC<br>NCC | INV<br>INV | End sequence with Step 16 |
| 24.12.8 | 3-color diamonds | Work row 1, then repeat row 2 | 1. CC<br>2. NCC | CC<br>CC | NCC<br>NCC | INV<br>INV | NCC<br>NCC | CC<br>CC | NCC<br>NCC | INV<br>INV | End sequence with Step 16 |

**Key:**  CC = Core change
NCC = No core change
INV = Inversion
—— = Skip step

# 24.13
## Embossed Nasca Diamonds with Four Colors

★ Adjust braid before working step

**Setting up**

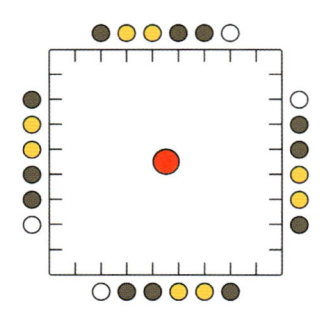

Braid ● 12 - 1 element
● 8 - 3 elements
○ 4 - 6 elements

Core ● 8 - 4 elements

24.13.1a Embossed four-color diamonds with fine outline threads.
Repeat Steps 1–24.

24.13.1b Embossed diamonds with plied groups of threads.

24.13.1c Using yarns of all the same thickness strengthens the dark outline but reduces the embossed effect.

Designs 24.13.1a and 24.13.1b are reproductions of a braid from the Nasca period (100–500 CE). Each element equals one thread. Varying the thread counts for each color gives this braid an embossed effect. The multiple groups of threads in the yellow, white, and red strands can be worked with or without plying them together. Design 24.13.1c uses strands of uniform thickness. See Table 3.1 on page 42.

*Working Instructions for Design 24.13.*

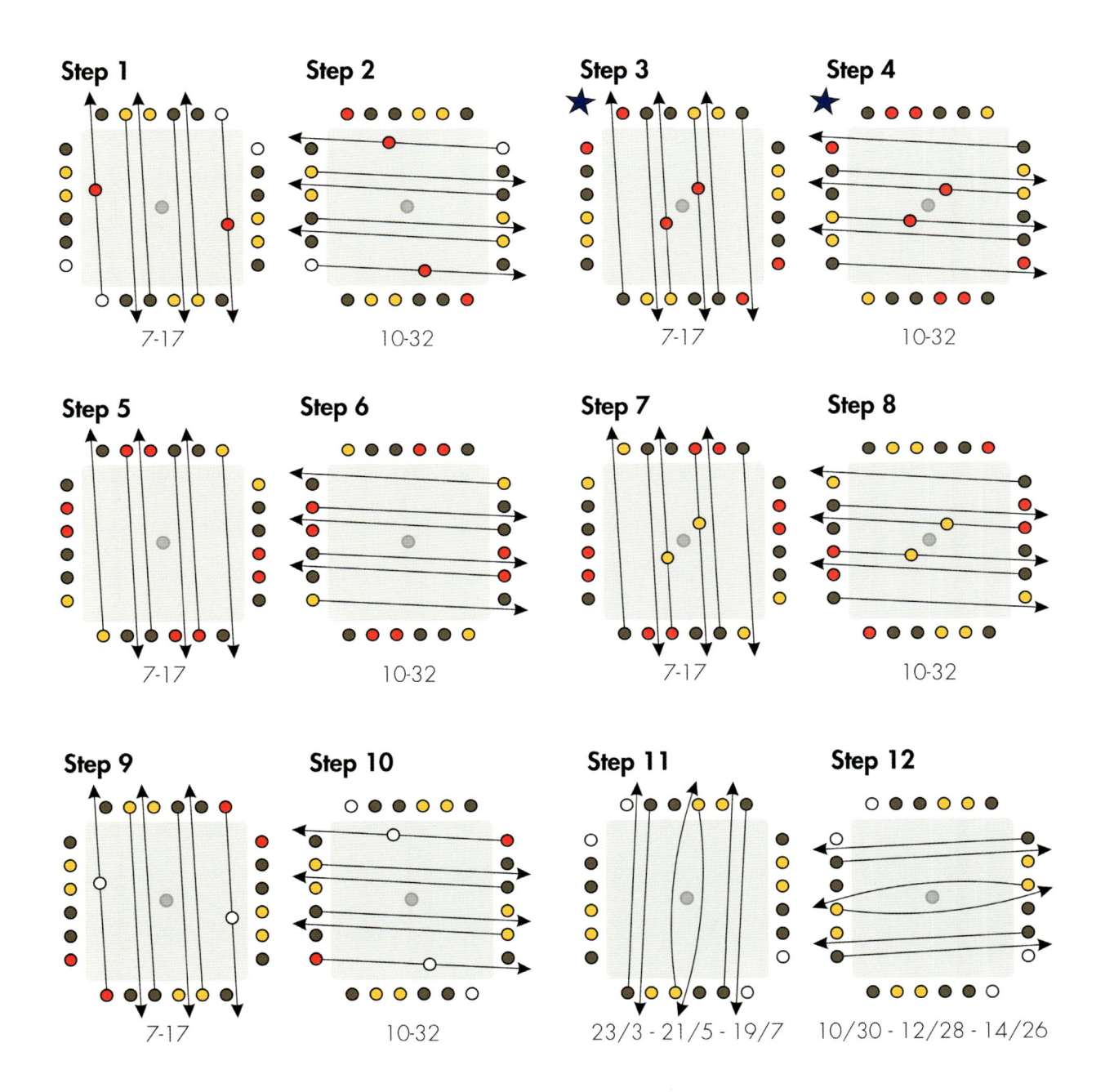

**Step 1**
7-17

**Step 2**
10-32

**Step 3**
7-17

**Step 4**
10-32

**Step 5**
7-17

**Step 6**
10-32

**Step 7**
7-17

**Step 8**
10-32

**Step 9**
7-17

**Step 10**
10-32

**Step 11**
23/3 - 21/5 - 19/7

**Step 12**
10/30 - 12/28 - 14/26

123

*Working Instructions for Design 24.13 continued.*

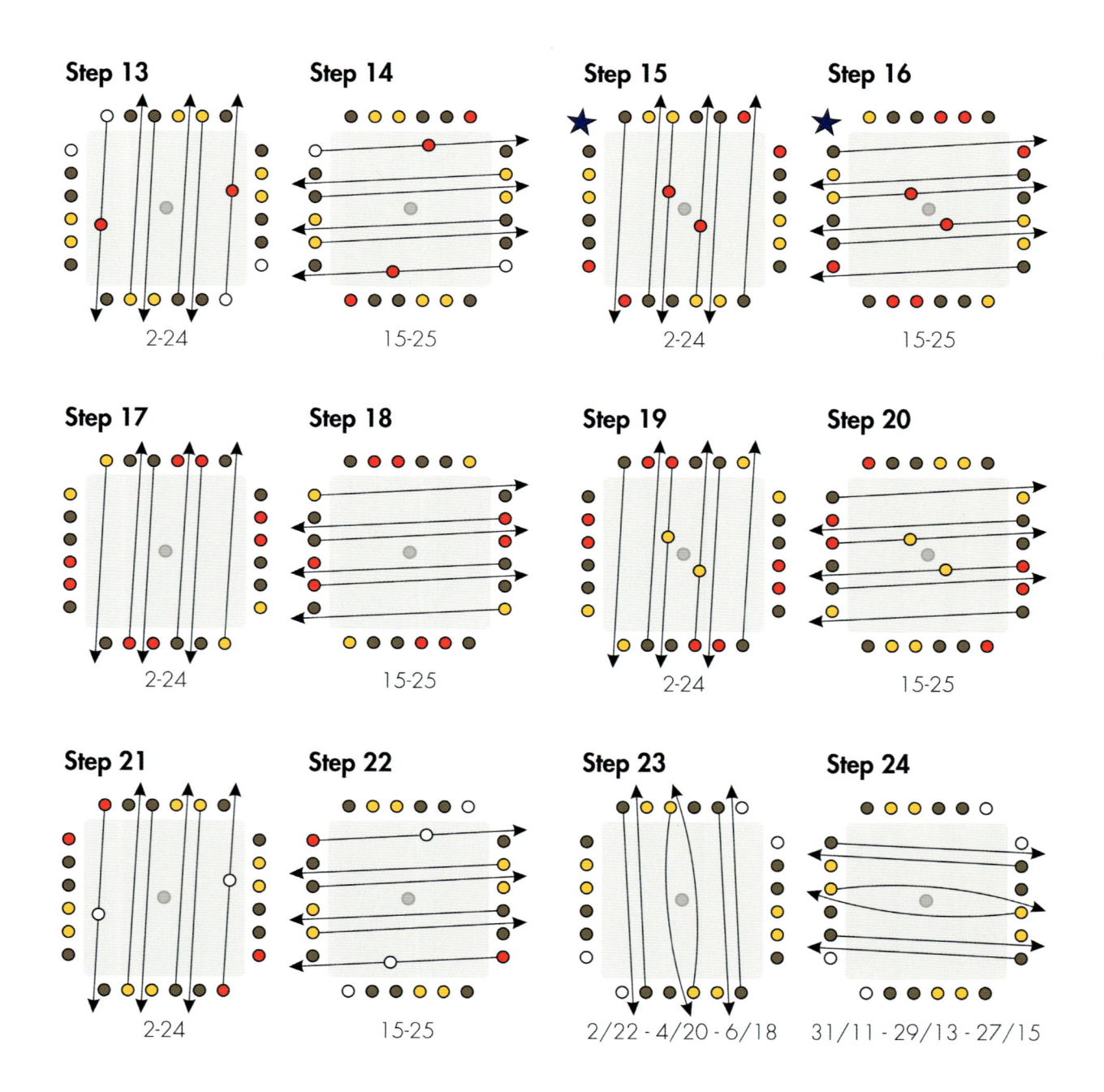

**Step 13**
2-24

**Step 14**
15-25

**Step 15**
2-24

**Step 16**
15-25

**Step 17**
2-24

**Step 18**
15-25

**Step 19**
2-24

**Step 20**
15-25

**Step 21**
2-24

**Step 22**
15-25

**Step 23**
2/22 - 4/20 - 6/18

**Step 24**
31/11 - 29/13 - 27/15

# 24.14
## Diamonds with Four Colors and a Core of 16

★ Adjust braid before working step

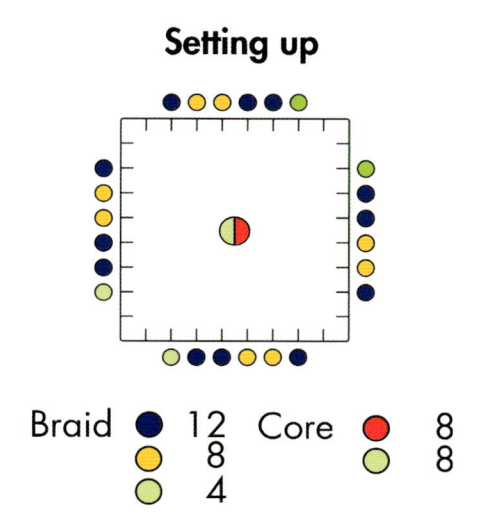

**Setting up**

Braid ● 12
● 8
○ 4

Core ● 8
○ 8

### 24.14.1a Four-color diamonds.
The side view of this braid has pink diamonds outlined in green with navy blue eyes.
Repeat Steps 1–20.

### 24.14.1b Variation with colors
of greater saturation and value contrast than in Design 24.14.1a.
Repeat Steps 1–20.

*Harness Set Up*
Arrange the colors on the harness arms as shown in figure 1. The four green threads near the center of the harness are placed in slots and should overhang in the direction shown in figure 2.

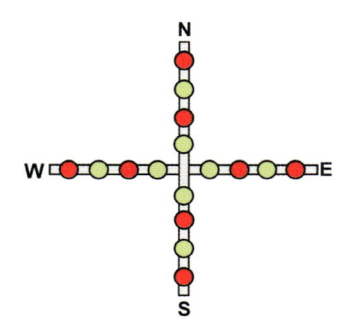

Figure 1 Core colors alternate on the harness in this 24-strand braid.

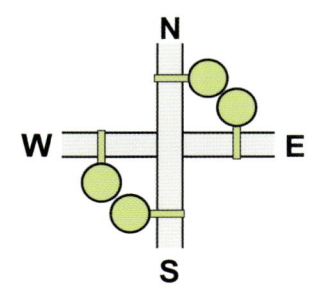

Figure 2 Initial position of the center core threads.

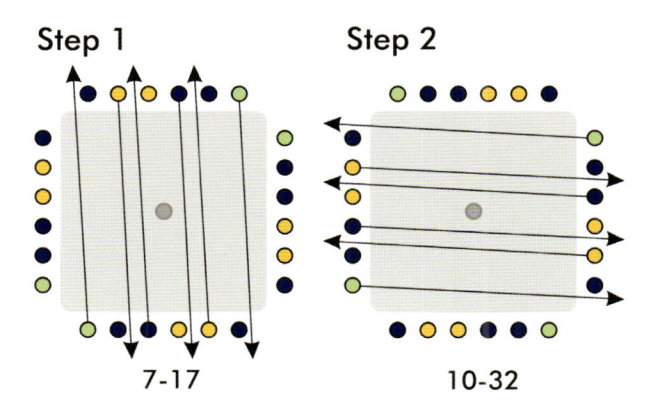

**Step 1**  7-17

**Step 2**  10-32

125

*Working Instructions for Design 24.14 continued.*

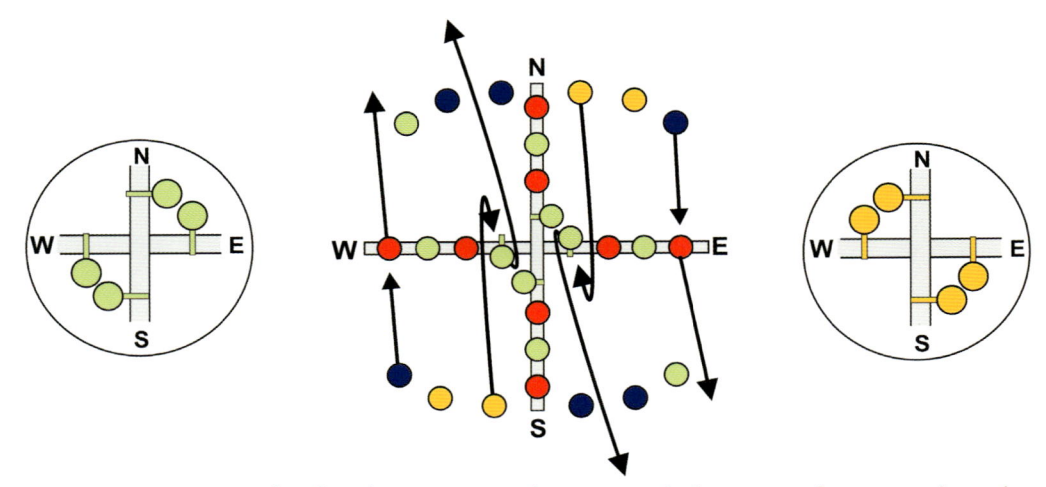

Figure 3  Details of making core exchanges with the center four core threads.

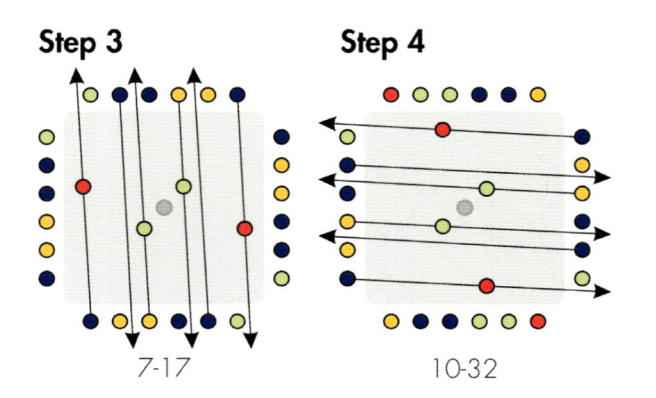

**Step 3**  7-17

**Step 4**  10-32

### Step 3  Core Exchange

(*Left*) Figure 3, left, shows the initial position of the four inner green threads on the harness. (*Center*) The threads are lifted off the harness and placed on the braiding stand in the direction shown. (*Right*) After the threads have been exchanged, the yellow threads will drape on the harness in the opposite direction.

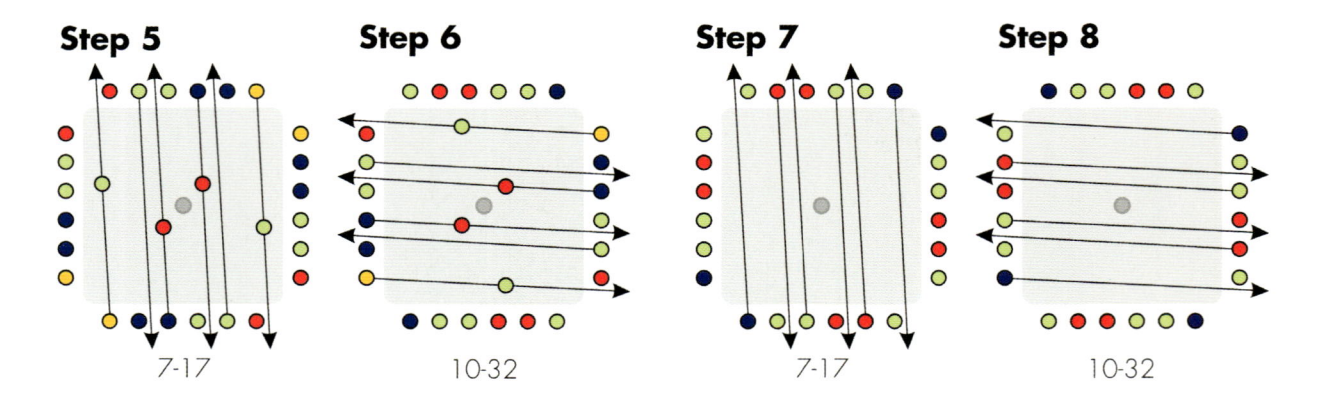

**Step 5**  7-17

**Step 6**  10-32

**Step 7**  7-17

**Step 8**  10-32

126

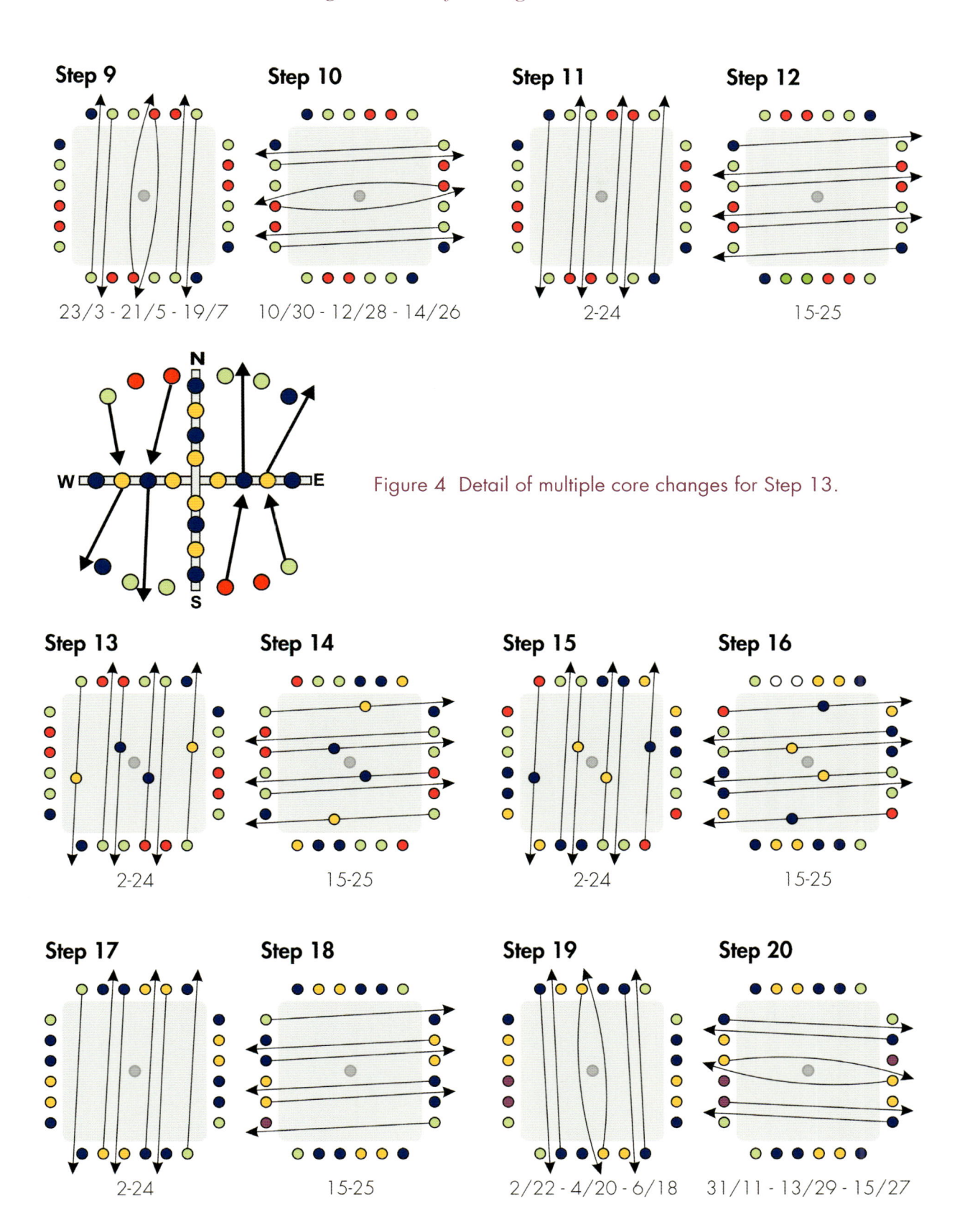

*Working Instructions for Design 24.14 continued.*

**Step 9**

23/3 - 21/5 - 19/7

**Step 10**

10/30 - 12/28 - 14/26

**Step 11**

2-24

**Step 12**

15-25

Figure 4  Detail of multiple core changes for Step 13.

**Step 13**

2-24

**Step 14**

15-25

**Step 15**

2-24

**Step 16**

15-25

**Step 17**

2-24

**Step 18**

15-25

**Step 19**

2/22 - 4/20 - 6/18

**Step 20**

31/11 - 13/29 - 15/27

127

# 24.15
## Diamond with Changing Eye Color (without a core)

This braid does not use a core; instead, diamond patterns with changing eye colors are created by repositioning the four upper threads at the commencement of each step. It is possible to make a braid with alternating eye colors along its length by working steps 1–17a, b, and c at the end of each color sequence. This braid can be also be worked with a single eye color.

**24.15.1b Diamonds with light eyes.**
Repeat Steps 1–17 to change to medium-colored eyes work steps 17a, b and c.

**24.15.1c Diamonds with medium-colored eyes.**
Repeat Steps 5–17 to make diamonds with medium-colored eyes.

**24.15.1a Diamonds with dark eyes.**
Repeat Steps 1–17 to change to a light-colored eyes work steps 17a, b and c.

**24.15.1d Diamonds with alternating eye colors.**

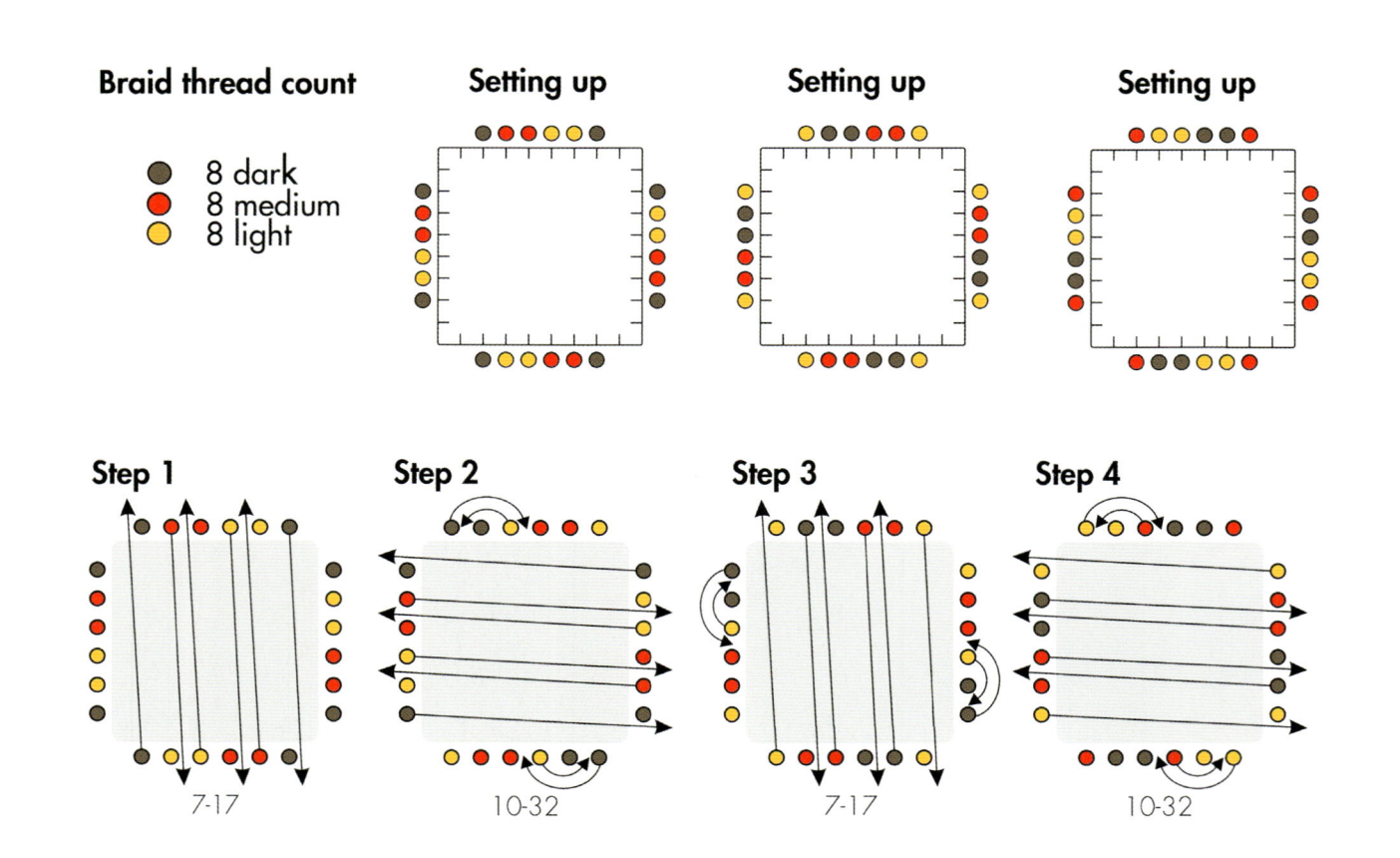

**Braid thread count**
- 8 dark
- 8 medium
- 8 light

Setting up

Setting up

Setting up

Step 1   7-17

Step 2   10-32

Step 3   7-17

Step 4   10-32

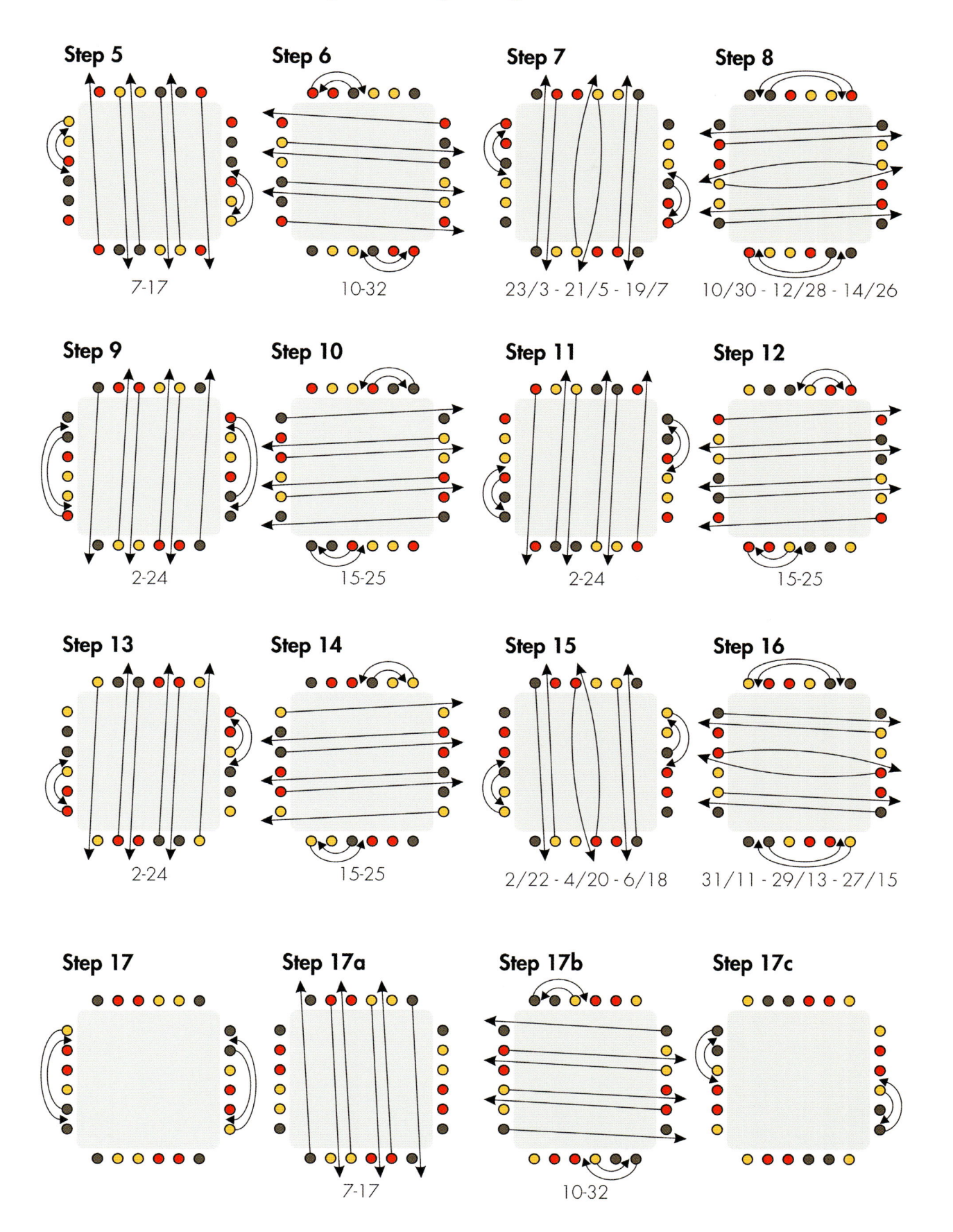

*Working Instructions for Design 24.15 continued.*

**Step 5**
7-17

**Step 6**
10-32

**Step 7**
23/3 - 21/5 - 19/7

**Step 8**
10/30 - 12/28 - 14/26

**Step 9**
2-24

**Step 10**
15-25

**Step 11**
2-24

**Step 12**
15-25

**Step 13**
2-24

**Step 14**
15-25

**Step 15**
2/22 - 4/20 - 6/18

**Step 16**
31/11 - 29/13 - 27/15

**Step 17**

**Step 17a**
7-17

**Step 17b**
10-32

**Step 17c**

# 32.1
## Elongated Diamonds (without core)

This 32-strand braid needs a card with 40 slots. The measurements for the card can be found in the Appendix.

**Setting up**

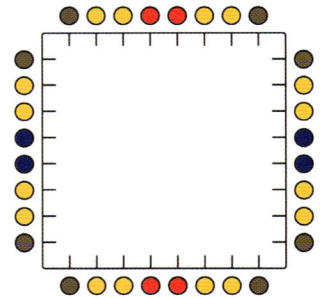

Braid ○ 16
● 8
● 4
● 4

32.1.1a (top)

32.1.1b (bottom)

Elongated four-color diamonds.

The way in which the colors are exchanged gives this braid its characteristic longer floats.

Repeat Steps 1–24.
Note that in Steps 5, 6, 9, 10, 17, 18, 21, and 22, the threads to be moved are upper threads, marked with the letter "U."

*Working Instructions for Design 32.1.*

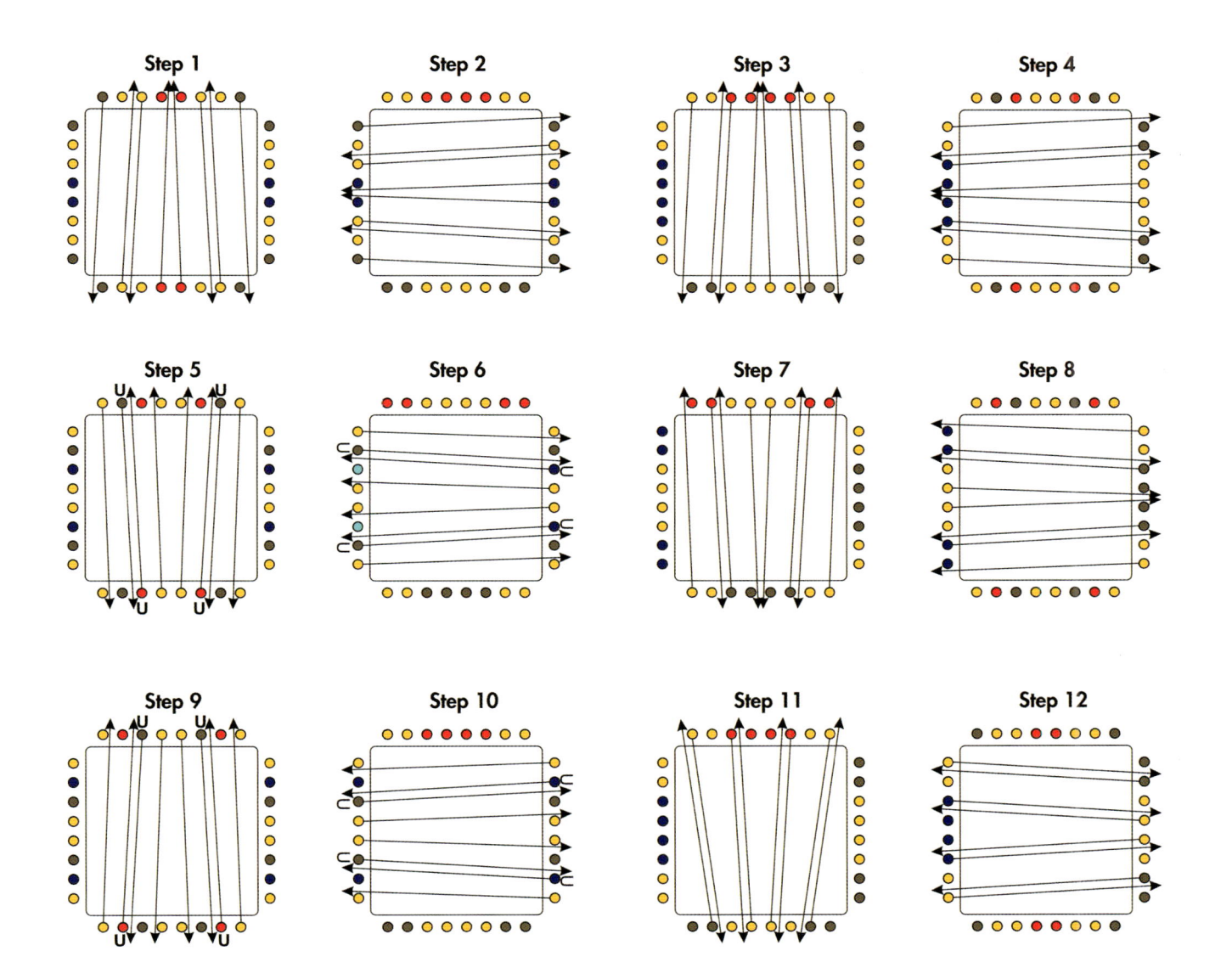

*Working Instructions for Design 32.1 continued.*

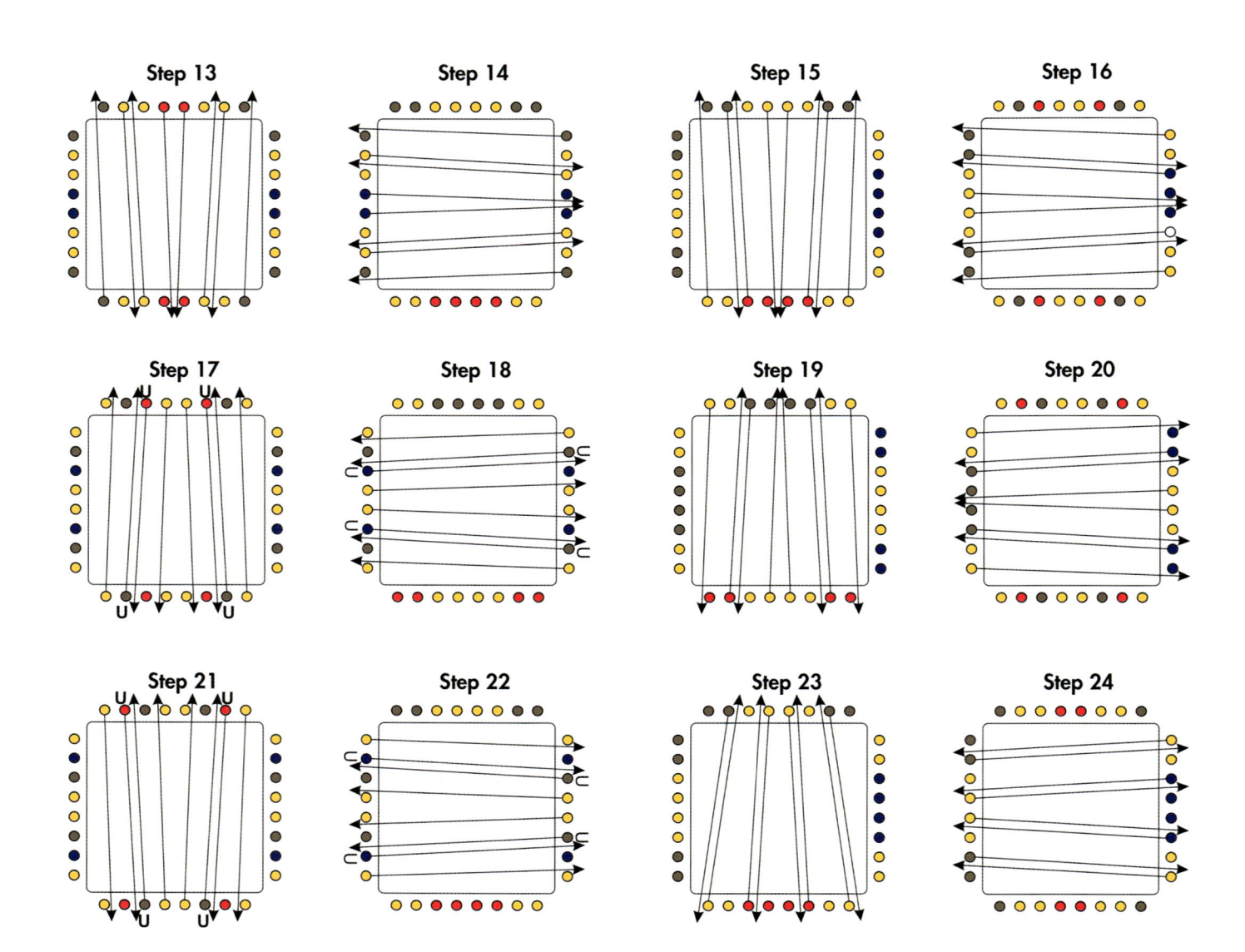

# 6. BEGINNINGS, ENDINGS, AND EMBELLISHMENTS

Andean sling makers use many specialized techniques that are both functional and decorative. Many of these, including braided loops and tasseled finishes, have many uses beyond sling making. This chapter gives instructions for these techniques as well as a few non-traditional methods of beginning and ending braids that we find helpful in certain applications. This dovetails with Chapter 7, *Making an Andean-style Sling,* which includes information on other useful techniques, such as how to add or remove threads to increase or decrease a braid's thickness, e.g., increase an 8-strand design to 16 strands, or decrease a 24-strand braid to 16 strands.

## Braided Loops

The purpose of incorporating a loop in one end of a sling is to prevent the sling from being released along with the missile. Slings can be started with a braided finger loop or a split loop can be worked near the end of a sling cord by dividing the warp into two smaller braids and rejoining them. (See figures 6.1 and 6.2.)

**FIGURE 6.2.**
Finger loop created near the start of the sling by dividing a 24-strand braid into two separate 12-strand herringbone braids and rejoining them. *Collection of R. Owen.*

### Starting a Braid with an 8-strand Loop on the Stand

To start a braid with an 8-strand loop, a short length of an 8-strand braid is made in the center of the warp. Next, eight bobbins are attached to the loose strands, and the warp is rearranged for a 16-strand braid. The first step of the 16-strand pattern joins the loop. The counterbalance weights are gradually increased as the braid is worked. The sample shown in figure 6.3 started with Design 8.1.1, the loop was closed, and the braid continued with Design 16.1.1.

**FIGURE 6.1.**
Sling started with a 12-strand loop. *Collection of R. Owen.*

**FIGURE 6.3.**
Spiral 16-strand braid started with an 8-strand braided loop.

*Preparing the warp and setting up for a loop.*

Prepare a warp for making a sample of the loop in figure 6.3. Tie the bindings while the warp is under tension. (Figure 6.4 shows the warp as it will appear after it is removed from the warping posts.)

- Set two warping posts 24" (61 cm) apart.

- Starting and ending at Post A, wind four of color 1 and four of color 2. (See Chapter 4, figure 4.2.)

- Temporarily mark the midpoint of the warp by tying a larkshead knot around the warp at Post B.

- To make a 2" (5 cm) loop, measure back 1.5" (4 cm) from the midpoint and tie bindings at points X and Y with a ¼" (6 mm) space between. (See figure 6.4.)

- Cut the warp from Post A.

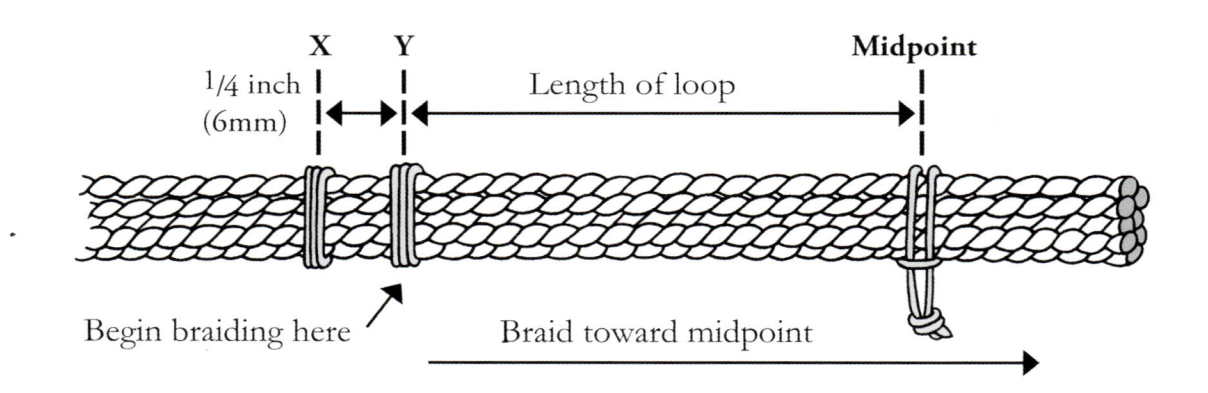

FIGURE 6.4.
Binding for beginning a braid with a loop.

**FIGURE 6.5.**
Make an S-hook loop by folding a 6" (15 cm) piece of binding string in half and tying the ends with an overhand knot.

Feed the bound end of the warp down through the center hole of the stand. Insert a double-pointed knitting needle between the bindings as shown in figure 4.8 and attach eight bobbins.

To make a strong finger loop, use a counterbalance weight slightly less than 50% of the total bobbin weight. Attach the counterbalance between the two bindings using an S-hook loop as shown in figures 6.5–6.7.

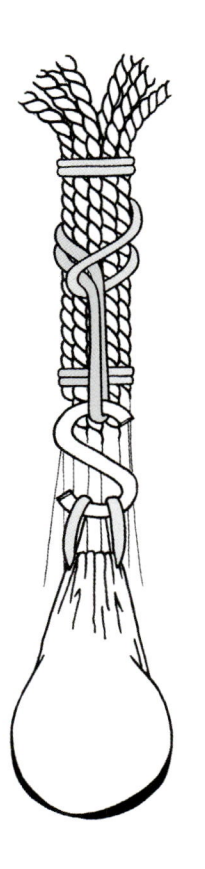

**FIGURE 6.6.**
Attach the loop between the bindings using a larkshead knot and attach the bag using an S-hook.

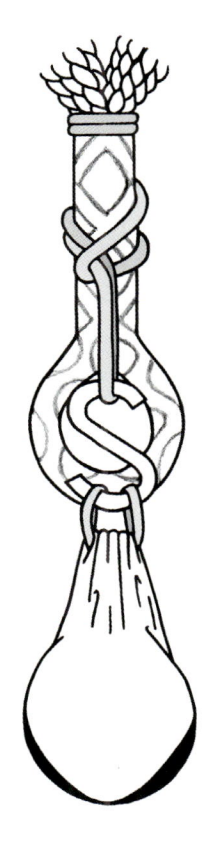

**FIGURE 6.7.**
Counterbalance bag suspended above a braided loop.

*Making the 8-strand braid for the loop.*

Arrange the threads for spiral Design 8.1.1 as seen in figure 6.8. Repeat the steps to make a 2" (5 cm) length of braid. To prevent the braid from unraveling while joining the loop, use a separate thread to tie it off on the underside of the stand. Insert a knitting needle over the warp from underneath the stand to stabilize the braid before removing the counterbalance bag. (See figure 6.9.)

## Design 8.1.1

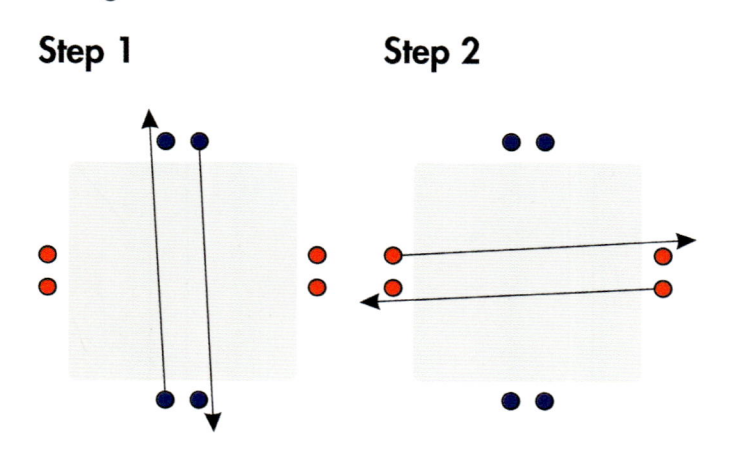

### Step 1       Step 2

**FIGURE 6.8.**
Repeat these two steps to make spiral Design 8.1.1.

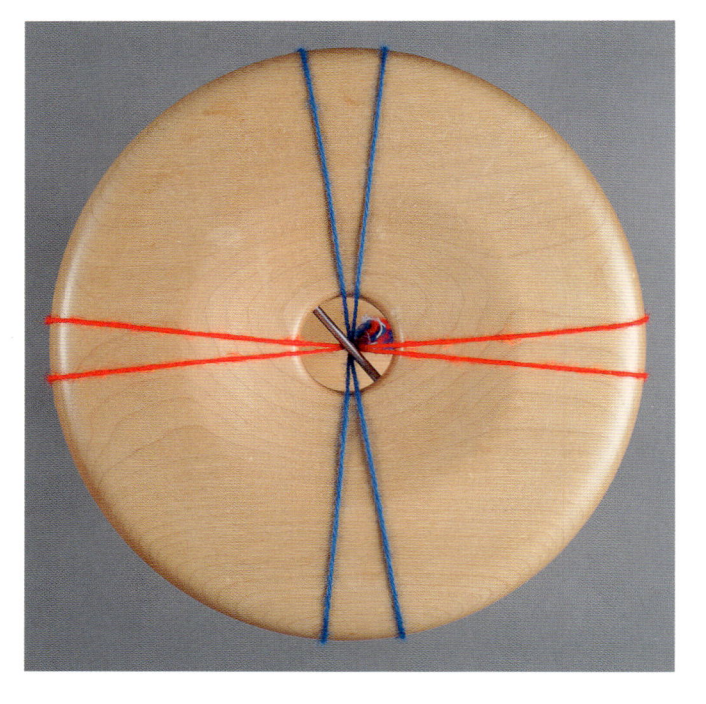

**FIGURE 6.9.**
Warp secured with a knitting needle.

Form the loop by reaching under the stand, folding the braid in half, and threading the loose warp up through the center hole. To keep the warp under tension while attaching the bobbins, reposition the knitting needle as shown in figure 6.10.

Untie the double bindings and attach the bobbins to the loose warp. So that the two ends of the braid interlace neatly, identify the lower threads and arrange them in the setup position for Design 16.1.1 in figure 6.12. Note that not all 8-strand braids will join with an identical interlacement.

**FIGURE 6.10.**
The knitting needle is positioned under the stand and over the braid while the loose end of the warp is arranged.

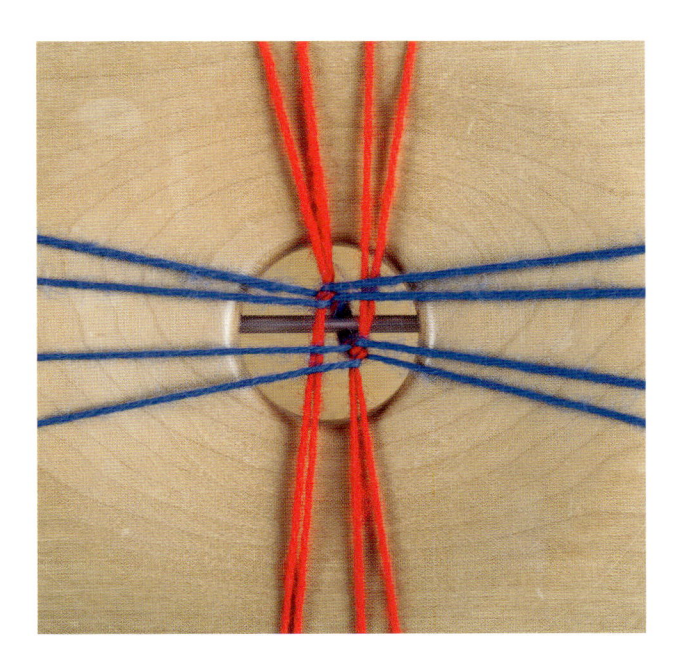

**FIGURE 6.11.**
The lower threads have been identified and positioned to begin making Design 16.1.1.

Referring to Design 16.1.1 in figure 6.12, work Step 1 with the red threads. This will pull the two ends of the loop together and secure the structure. Braid the first four steps *without a counterbalance*, and then reattach the S-hook loop above the braided loop as shown in figure 6.7. Gradually increase the counterbalance weight a few ounces at a time as you work the next few steps.

### Starting an 8-strand Loop on the Card

The procedure for making a loop on the card follows the same process as making one on the stand. The completed loop braid with a diamond pattern is shown in figure 6.13. Prepare a warp for Design 8.2.2 following the instructions for the stand, then:

- Braid a 2" (5 cm) length, ignoring the instructions regarding the counterbalance bag

- Bring the braided length up through the center hole and release the temporary bindings as seen in figure 6.14

- Identify and rearrange the lower threads for Design 16.3.1 seen in figure 6.20

## Design 16.1.1

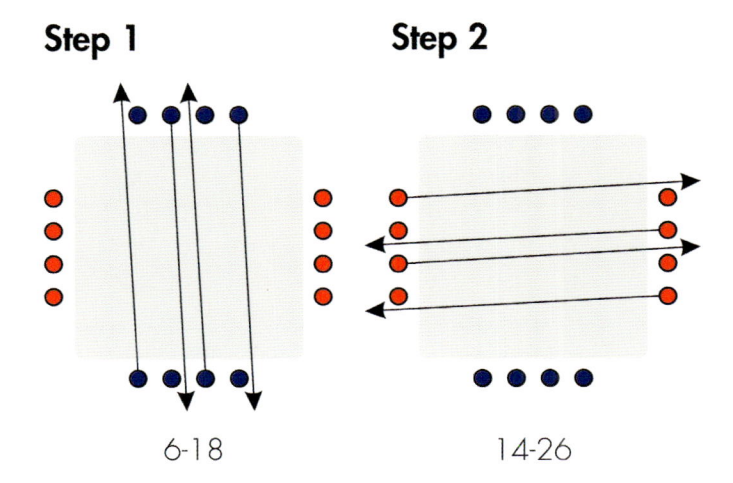

**Step 1**       **Step 2**

6-18              14-26

**FIGURE 6.12.**
Repeat these two steps to make spiral Design 16.1.1.

**FIGURE 6.13.**
Loop braid using Designs 8.2.2 and 16.3.1.

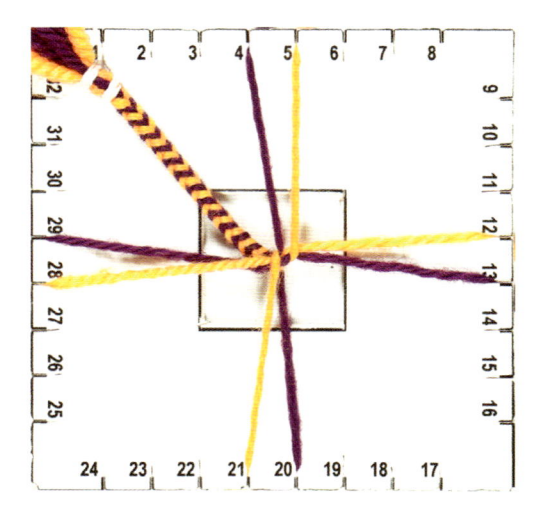

Figure 6.16 shows a variety of braids that can be designed by incorporating various 8- and 16-strand patterns. Loop braids have many uses including connecting jewelry elements to a necklace as in the pendant in figure 6.17. To make a 24-strand braid from an 8-strand braided loop, make the loop over a core of eight strands. See Chapter 7 for additional methods for increasing or decreasing a braid from one size to another.

**FIGURE 6.14.**
An 8-strand braid has been brought to the top side of the card in preparation for joining to form the loop.

**FIGURE 6.16.**
A variety of loops worked in different 8-strand patterns that transition to 16-strand braids.

## Design 16.3.1 Card Moves

| Clockwise | | Anticlockwise | |
|---|---|---|---|
| 6-18 | 14-26 | 3-23 | 11-31 |
| 5-6 | 13-14 | 4-3 | 12-11 |
| 20-5 | 28-13 | 21-4 | 29-12 |
| 4-20 | 12-28 | 5-21 | 13-29 |
| 3-4 | 11-12 | 6-5 | 14-13 |
| 22-3 | 30-11 | 19-6 | 27-14 |

On completing each column, reposition the threads to their set-up slots.

**FIGURE 6.15.**
Card moves for Design 16.3.1.

**FIGURE 6.17.**
Braided loop used to attach a pendant to a necklace.

**FIGURE 6.18.**
Dog lead made with split loop for the handle.

## Making a Split Loop

Braids of 16 or 24 strands can be divided into two separate braids that are rejoined to create a loop as is often seen in slings. (See figure 6.2.) The dog lead in figure 6.18 was begun in the middle of the warp with Design 8.2.2. (Tip: reduce the counterbalance to less than 50% of the bobbin weight in order to make a firm loop that will wear well.) The clip was slipped over the loose ends of the warp and the braid was joined with the method explained above in *Closing the loop and beginning the 16-strand braid*. A length of Design 16.3.1 was made. To create the handle, the braid was divided into two 8-strand braids and then rejoined to close the loop. (See Steps 1–5 below.) The lead was finished with a simple tassel.

Card braiders will follow the same steps using corner slots to hold the passive braid threads until it is their turn to work.

**FIGURE 6.19.**
Design 16.3.1 split into two braids of Design 8.2.2 and rejoined.

Note: It will not always be possible to avoid occasional floats when transitioning between 16-strand and 8-strand designs. Pay attention to the positions of upper and lower threads before braiding a split or combined section and correct positioning when necessary, as shown in Move 1. To make a sample of a split loop like the one in figure 6.19 follow Steps 1–6 in Figures 6.20–6.24.

## Step 1
6-18

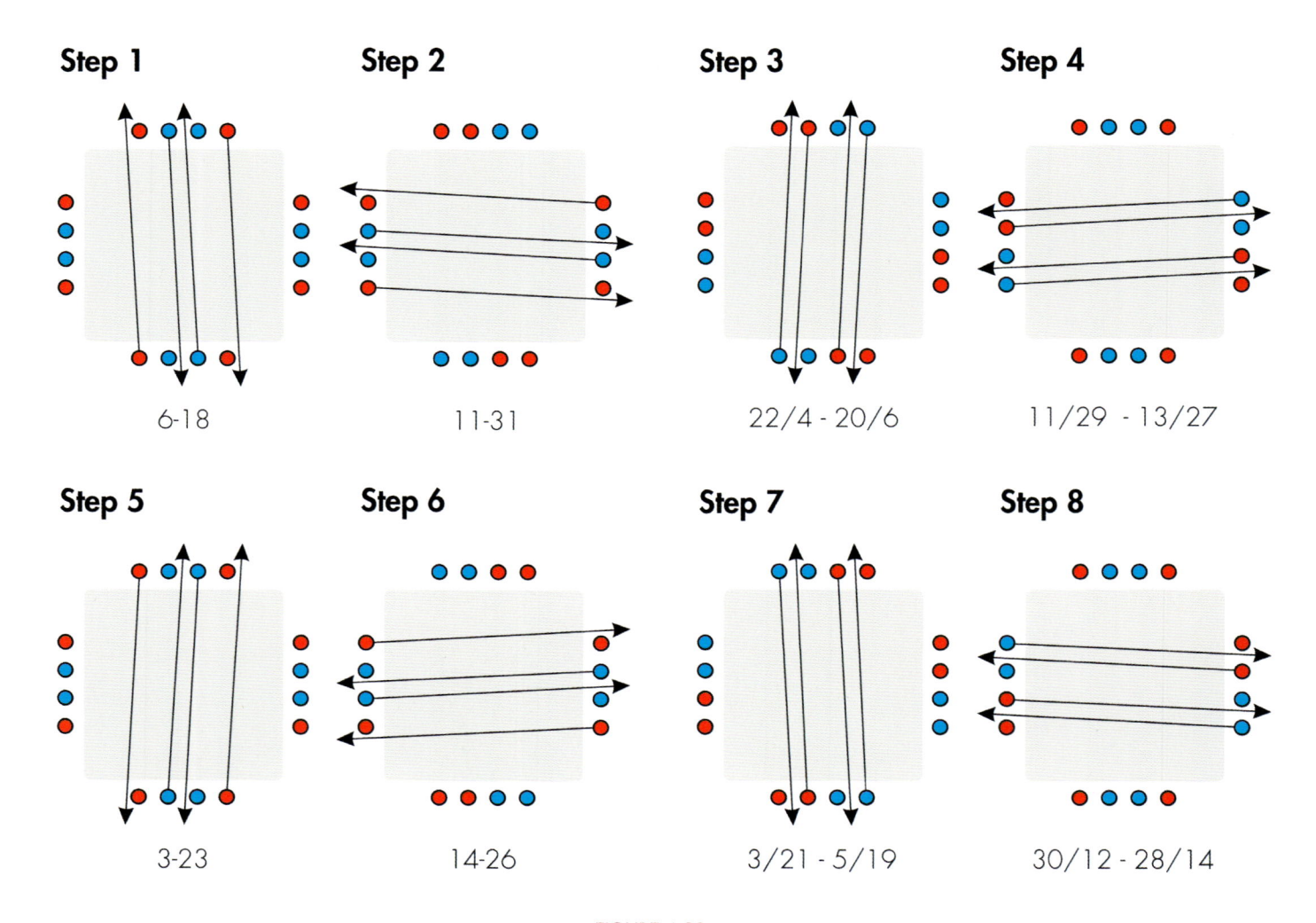

## Step 2
11-31

## Step 3
22/4 - 20/6

## Step 4
11/29 - 13/27

## Step 5
3-23

## Step 6
14-26

## Step 7
3/21 - 5/19

## Step 8
30/12 - 28/14

**FIGURE 6.20.**

STEP 1.
Begin by making a length of Design 16.3.1, repeating Steps 1–8.

**Leg 1**

**Move 1**   **Move 2**

**Design 8.2.2**

**Step 1**   **Step 2**

FIGURE 6.21.

STEP 2.
Divide the threads for the first 8-strand braid
by making *Moves 1 and 2* as follows:

*Move 1.* Slide the N&S pairs to the corners.
Slide the W pairs to the corners.
Reposition the two S threads to prevent a float by lifting
the blue thread over the red.

*Move 2.* Bring the top pair of threads
at the E across to the W.
The light red and blue threads will be inactive while
the first 8-strand braid is made.

FIGURE 6.22.

STEP 3.
Start the braid by repeating Steps 1 and 2.
Depending on the length of the loop,
it may be necessary to work first one braid for an
inch or two, and then the other.

**Leg 2**
**Move 3**  **Move 4**  **Move 5**  **Move 6**

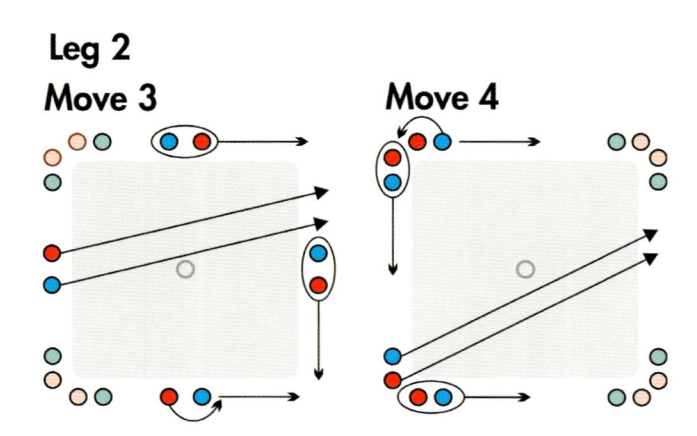

FIGURE 6.23.

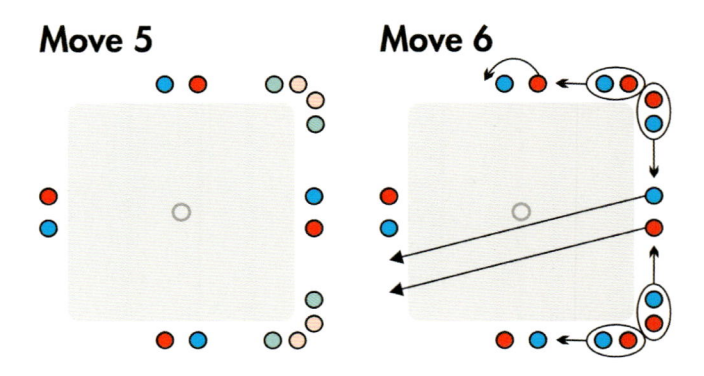

FIGURE 6.24.

STEP 4.
To begin the second leg of the braid, place the first group of threads to the E side of the stand.

Move 3. Slide the circled N&E threads to the corners. Lift the S red thread over the blue and slide to the corner. Move the W pair to the E and slide to the upper corner.

Move 4. Move the passive threads into position by sliding the upper and S pairs to the central spots. Before positioning the N pair, lift the blue thread over the red. Center the remaining W pair.

STEP 5.
With the threads in the setting up position shown in Move 5, make a length of 8-strand braid to match the first leg. End with the colors in the correct setup position for the first step of Design 8.2.2 in figure 6.22. Work first one leg and then the other until they are ready to be joined.

Move 6. Lift the red N thread over the blue as seen in Move 6. Move the E pair to the W. Slide the remaining pairs into position ready to make Step 1 of Design 16.3. (See figure 6.20.) Adjust uppers and lowers as necessary. Braid Steps 1 and 2 of Design 16.3 to close the loop. Carefully tighten the N/S threads. Work Step 3 and carefully tighten the E/W threads. Continue braiding for the desired length.

## Starting a 16-strand Braid with a Blunt End, Method B

Method A for beginning a braid with a blunt end using a larkshead knot is described in Chapter 4, *Preparing the Warp*. Method B offers a more precise way to create a blunt end that has a neater beginning. It is useful in applications where floats or uneven stitches are undesirable as when jewelry findings are embedded at the beginning of the interlacement.

Stand and card braiders will prepare the warp in the same way. Calculate the warp length for your project (including take-up) and double this number. Wind the warp and cut the warp at post A. To make a 9" (23 cm) sample of the braid shown in figure 6.25, wind a 1 yd. (1 m) warp for each of the four colors shown in the set-up diagram in figure 6.25 (or substitute colors of your choice). The warps will be added to the structure in pairs. (See figures 6.26 and 6.27.) Card braiders will proceed to "*Layering the Warp*." Stand braiders will first attach bobbins to both ends of each warp as described next.

### Attaching Bobbins to Both Ends of a Warp

To wind equal amounts of warp onto each bobbin, temporarily anchor each strand before attaching the bobbins as follows:

- Make a slipknot in the center of the first warp strand

- Place the loop over a T-pin anchored in a carpet, onto a warping post or other stationary object

- Add a bobbin to each end of the warp. Adjust the lengths to 10" each

- Leaving the first pair in place, add bobbins to the next pair

- Remove the most recent pair, pull out the slipknot, and set the warp aside. Continue until all 16 bobbins have been attached, measuring each against the first pair

Making a Sample of Design 16.1.3 with a Blunt-end Beginning

Referring to the color set up seen in figure 6.25:

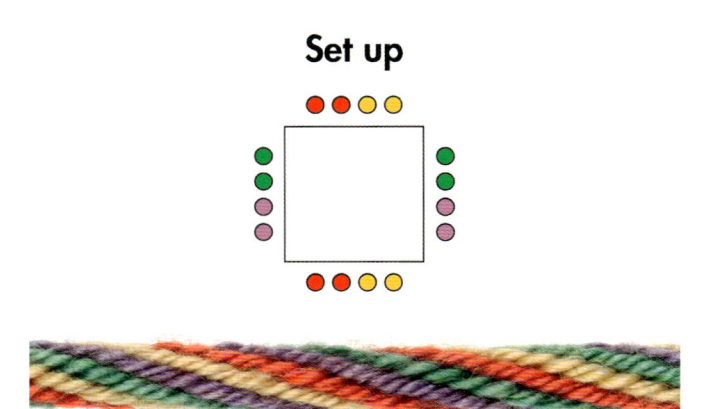

**FIGURE 6.25.**
Four-color spiral braid, Design 16.1.3.

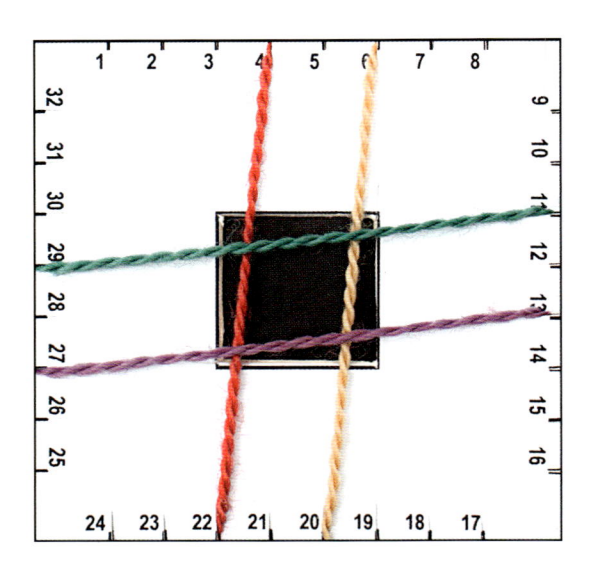

**FIGURE 6.26.**
The lower threads are placed first to the N/S, then to the E/W.

**Stand:** Lay the threads across the stand in the color order and sequence in which they will be moved, i.e., the "lower" threads of the structure are placed first to the N/S, then to the E/W, followed by the "uppers" to the N/S and E/W. Follow the order shown in figures 6.26 and 6.27.

**Card:** Place the middle of each warp over the center hole and lay the ends into the slots in the sequence described below and shown in figures 6.26 and 6.27:

- Place the lower threads in N/S slots 4-22 and 6-20
- Place the lower threads in E/W slots 11-29 and 13-27
- Place the upper threads in N/S slots 3-21 and 5-19
- Place the upper threads in E/W slots 12-30 and 14-28

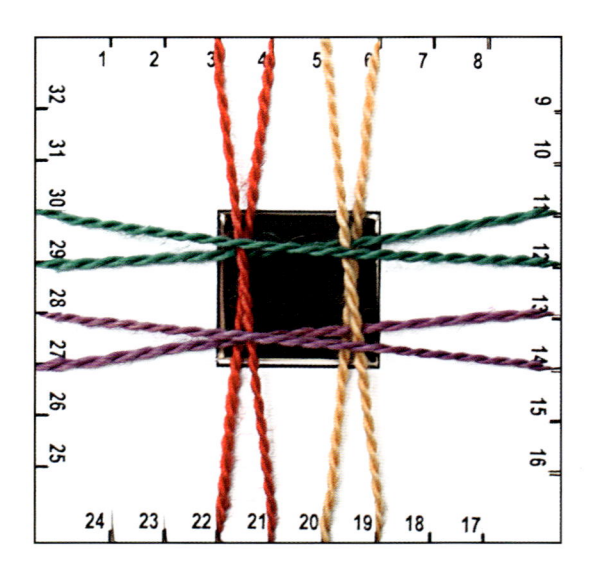

**FIGURE 6.27.**
The upper threads are laid in on the diagonal on top of the lower threads.

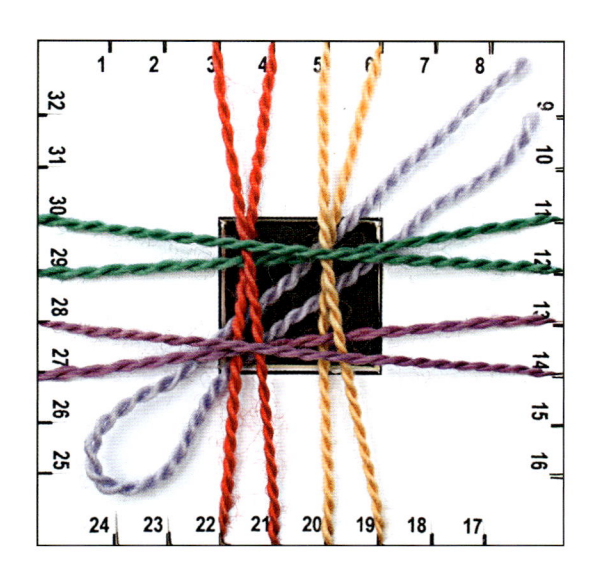

**FIGURE 6.28.**
Insert the loop of the larkshead tie under all of the threads.

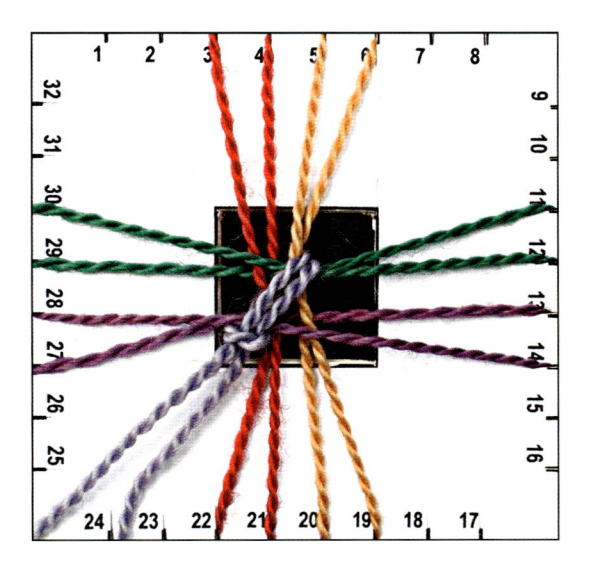

**FIGURE 6.29.**
Bring the two loose ends over the threads and bring through the loop to form a larkshead knot. Pass the loose ends down through the center hole and pull tight.

For both stand and card methods, secure the warp as follows:

- Cut a 7" (18 cm) length of yarn, fold in half, and insert under the assembled warp as shown in figure 6.28.

- Bring the two loose ends over the warp and pass them through the loop to form a larkshead knot. (See figure 6.29.)

- Pull the knot tight and drop the two loose ends down through the hole.

- For *stand braiding* tie an overhand knot in the loose ends ½" (2.5 cm) from the larkshead.

- For *card braiding* tie an overhand knot a finger's length from the larkshead to make a loop for holding the braid under the card.

To prevent loose stitches in the start of the braid, work the first two steps without tugging on the threads. Before making Step 3, carefully tighten the N/S threads and then work the step. Before making Step 4, carefully tighten the E/W threads and then work the step. Check the tension before working the next two steps to make sure there are no loose stitches.

## Starting a 24-strand Braid with a Blunt End

Design 24.1.1c, shown in figure 6.30, is used to illustrate starting a 24-strand braid with a blunt end. To arrange the threads for the stand and the card follow the same procedure as for the 16-strand braid. Wind a warp using the color set up seen in figure 6.30 or substitute colors of your choice.

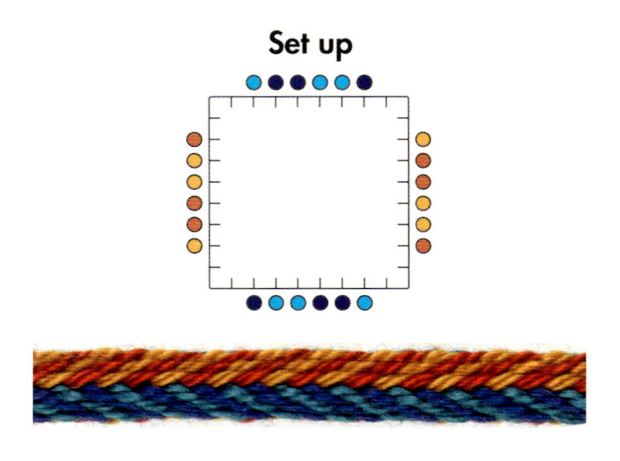

**FIGURE 6.30.**
Four-color set up for Design 24.1.1c.

**Stand Method:** In the same manner as setting up the braid in figures 6.26 and 6.27, place "lower" and "upper" threads to the N/S and then E/W as follows:

- Place the three lower N/S warps, then three lower E/W warps
- Place the three upper N/S warps, and last place three upper E/W warps

When all the threads are in place, gather and secure the warps with a larkshead knot as shown in figures 6.28 and 6.29. Tie an overhand knot in the ends and insert the knitting needle. Attach the counterbalance bag and begin braiding. If the bobbins are not all the same length, braid the first eight steps and then adjust.

**Card Method**: Place the warps in the slots in the following order:

- First three lower N/S threads in slots 3-23, 5-21, and 7-19
- Three lower E/W threads in slots 10-30, 12-28, and 14-26
- Three upper N/S threads in slots 2-22, 4-20, and 6-18
- Three upper E/W threads in 11-31, 13-29, and 15-27

When all the threads are in place, secure with a larkshead knot and tie an overhand knot as before. Be careful not to pull tightly as you braid the first four steps. Then stop and gently snug up any loose stitches.

## Starting a 24-strand Braid with a Hollow-braid Wrapping

Wrapping, as seen at the beginning of the braid in figure 6.31, is an attractive technique for beginning and/or ending braids used in jewelry. It creates a firm base onto which findings can be attached. Terry developed this spiral wrapping technique while experimenting with ways to "trap" fly-away Japanese silk used in necklace cords. It could also be used to create additional texture or pattern in a braid.

Figures 6.32–6.33 give the steps for using a 12-strand hollow braid to wrap a 24-strand braid. The example shown does not have a core; however, the technique can also be used for braids with cores. (This method is not recommended for card braiding since it is easier to work with the core suspended.)

The process begins by working the hollow braid in color A around the core of color B. When the desired length of color A has been worked, color B is introduced and the two-color pattern begins.

**FIGURE 6.31.**
Braid started with a wrapped blunt end.

## Setting up a Hollow Braid

This set up creates a blunt beginning with no floats. To make a sample braid, prepare a warp by cutting six lengths each of two colors, 39" (1 m) long. Attach bobbins to both ends of each warp as described in *Starting a 24-strand Braid with a Blunt End*.

*Working clockwise*, and referring to figure 6.32, lay the pairs on the stand as follows:

- Beginning with the N/S pair, position pairs of white lower threads (indicated by the dotted lines) as shown in Step 1, figure 6.32

- Next, lay in the white upper threads (indicated by the solid lines), beginning with the N/S pair shown in Step 1

- Place the red core threads between the white threads as shown in Step 2

- Secure the warp with a larkshead knot

- Attach a counterbalance and lift the red threads onto the harness

### Working a Hollow Braid over a Core

To work the white hollow braid over the red core, repeat Steps 3 and 4 in figure 6.33, ending with Step 4.

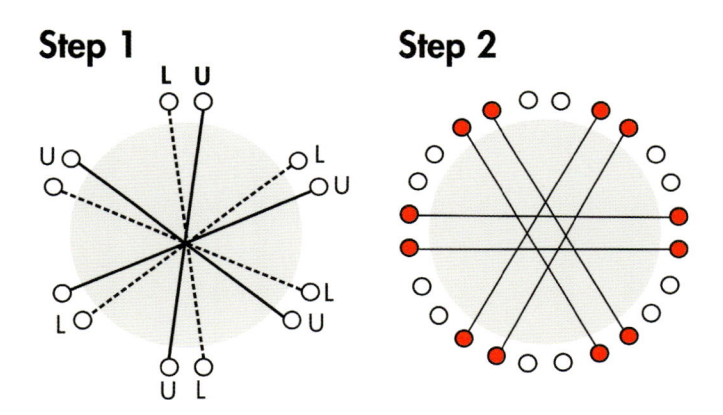

FIGURE 6.32.
Arrange the white threads, lowers first, then uppers. Then add the red core threads.

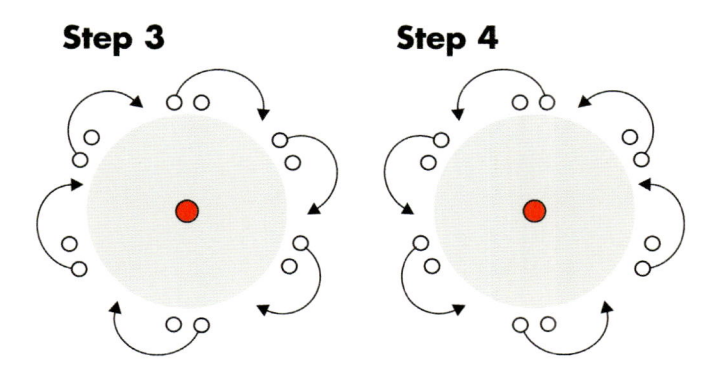

FIGURE 6.33.
Wrap the beginning of the braid by repeating Steps 3 and 4.

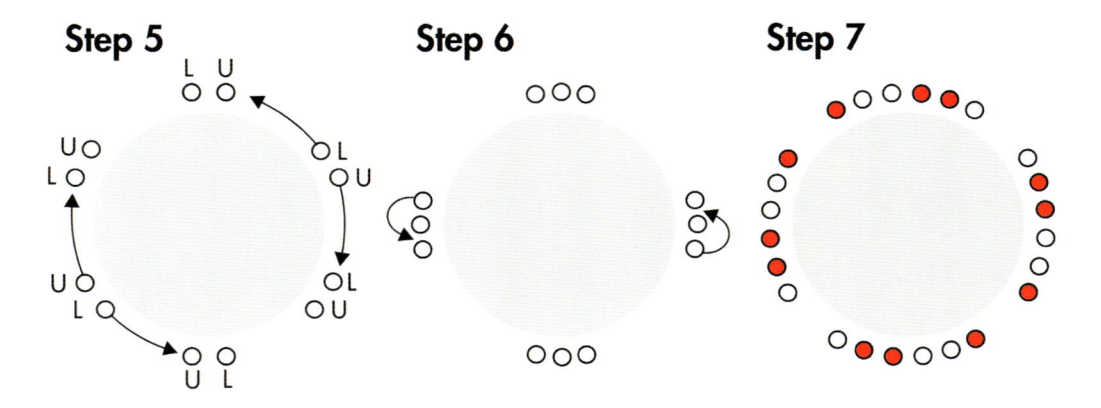

**FIGURE 6.34.**
Steps for making the transition between hollow braid and chevron or diamond braid structures.

## Transitioning to a Chevron or Diamond Pattern

After the desired length of hollow braid has been made, the threads must be positioned carefully to make a neat transition to the chevron/diamond pattern. To change from hollow braid to core braid structure, rearrange the lower and upper white threads in groups of three as shown in Step 5, figure 6.34. To avoid unwanted floats, make the adjustment shown in Step 6 by moving the designated bobbins. Last, bring the red threads down from the harness and place them between the white threads in the positions shown in Step 7. To make a length of chevrons, repeat the two steps shown in figure 6.35.

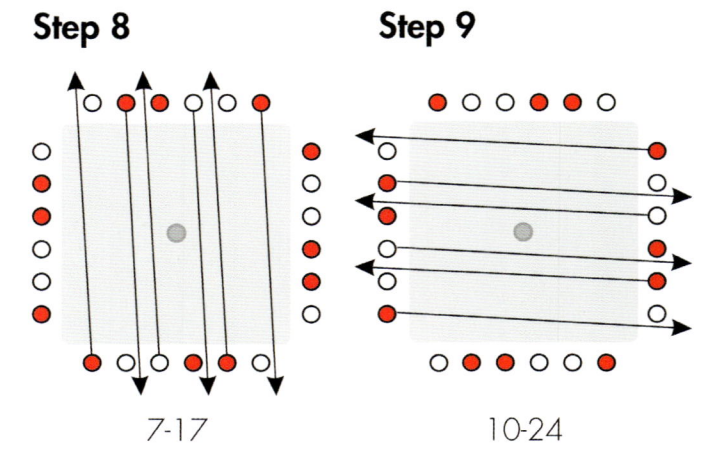

**FIGURE 6.35.**
Repeat Steps 8 and 9 to make a chevron pattern.

## Stitched Finishes

A variety of stitching techniques are used on Andean braids. These can be functional— used to reinforce the edge of a sling cradle, or decorative—when applied to making tassels.

### Stitched "Beads"

Andean braid wrappings that look like beads are often stitched in rows above tassels. Rodrick has adapted this technique for jewelry.[1] Silk textile beads as seen on the necklace in figures 6.36 and 6.37, can be added to a braid at any point along its length. They can serve as both an embellishment, as well as an effective way to keep beads or pendants from sliding out of place. Make sure to sample this method with different types of yarn, as it takes practice to master sewing with an even tension.

To make a bead, cut a length of yarn sufficient to complete the bead without adding a new strand. Sample with an 18" (45 cm) length and adjust as you get a feel for the kinds of threads you use. Thread a blunt-pointed needle and sew through the braid to anchor the thread. (A size 18 tapestry needle works with average-sized threads.)

• • • • • • • • • • • • • • • • • • • • • • • • • • • • • • • • • • • • • • • • • • • • • •

Tip: To thread yarn into a tapestry needle, fold 2" (5 cm) at the end of the strand. Place the needle in the fold. Pinch the yarn and needle between the thumb and index fingers of your non-dominant hand. This flattens the yarn. Holding the needle in your dominant hand, pull the needle out. Then push the flattened yarn into the eye of the needle. With a little practice, this is quick. An alternative is to use a needle threader or loop of thin cotton.

• • • • • • • • • • • • • • • • • • • • • • • • • • • • • • • • • • • • • • • • • • • • • •

The profile of the wrapping should resemble a pyramid. This creates the foundation over which the bead will be made. (See figure 6.38.) Wrap the thread firmly (but not tightly) around the braid four turns. Do not overlap. Wrap three turns on top of the initial four turns, carefully placing each wrap "in the ditch" between the threads below. Add two more wraps to the last level.

Stitch over the initial wrappings to create the bead. Work with an even tension, keeping the bead stitches close together. Finally, sew the end into the braid and cut. (If it is difficult to slide the needle under the wrapping, switch to a thinner needle or undo, and wrap a little less tightly.)

**FIGURE 6.36.**
The Andean-style stitched beads in this silk necklace prevent the Turkmen pendent and the glass and silver beads from sliding out of place.

**FIGURE 6.37.**
Detail of stitched beads.

**FIGURE 6.38.**
A wrapped foundation is built up in layers and then stitched over.

## Inserted Tassels with Two Styles of Binding

Embellishing festival braids with complex tassels or pompoms is part of the Andean textile tradition. The tassel skirt is often inserted into an existing braid. The braid gives the skirt support so that even soft yarns have a nice full appearance. (Different from European tassel skirts that are often made over a wooden form.) Tassels can be spaced closely together, or spaced to show off the supporting braid. (See figures 6.39 and 6.40.)

Eleuteria Nuñez Huamani, of Ayacucho, Peru, made the beautiful embellished braids seen in figure 6.39. She started these by braiding a loop (sometimes using doubled stands), making a length of 24-strand core braid, and dividing this into two or three 4- or 8-strand braids. Her method for binding the tassels is described in *Tassel Binding Method 1*.

The colorful tassels shown in figure 6.40 embellish a long Andean belt. These are sewn through 24- and 8-strand braids and are bound with *Tassel Binding Method 2*. (See figures 6.41 and 6.42.)

The sample tassels in figures 6.41 and 6.42 were made by sewing multiple strands of knitting yarn through a braid made of alpaca. The yarns were smoothed down the sides of the braid, and then bound with stripes of a thinner, contrasting thread using *Tassel Binding Methods 1 and 2*.

**FIGURE 6.39.**
Eleuteria Nuñez Huamani makes exquisite 24-strand core braids using the traditional hand braiding method finishing them with tassels or pompoms. They measure about 6¾" long × 5/16" diameter

**FIGURE 6.41.**
Tassel Binding 1 is wrapped snugly around the skirt.

**FIGURE 6.40.**
Tassels attached to a long Andean belt using Tassel Binding Method 2. *Courtesy of Barbara Wolff.*

**FIGURE 6.42.**
Tassel Binding 2 is stitched around bundles of skirt threads.

151

*Inserting a tassel skirt.*

Cut the tassel yarn twice the length of the finished tassel plus 1½" (4 cm) ease for trimming. The tassels shown in figures 6.41 and 6.42 had skirt yarns cut at 8" (20 cm). The finished length after trimming was 2½" (6 cm). Thread several strands together in a tapestry needle. (Use a needle threader or a loop of cotton thread to stuff a lot of yarn into a size 18 tapestry needle.) It may take some experimenting to find the optimum number of strands. Stitch through the braid from N to S, leaving equal amounts of skirt on each side. Continue sewing in skirt yarns around the braid (E/W, NE/SW, and NW/SE), smoothing down the strands along the braid, and checking to see that the braid is well covered.

*Tassel binding method 1.*

This binding is used in Eleuteria's tassels. (See figure 6.39.) The sample in figure 6.41 was tied off with embroidery thread in blue/yellow/blue as follows:

- Cut a 12" (30 cm) length of a thin thread and fold in half

- About ½" (13 mm) down from the top of the tassel, wrap the loop around the tassel and pull the ends through to make a larkshead knot. Pull tight. Thread a tapestry needle with the loose ends

- Wrap a second time. Sew into the loop of the larkshead knot and down into the braid towards the tassel skirt

- Repeat for binding each color

- Trim the tassel to the desired length

*Tassel binding method 2.*

Tassel Binding 2, seen in Figure 6.42, is not wrapped. Instead, groups of skirt threads are stitched together using a split-stitch that wraps around bundles of skirt threads, *while not stitching into the braid*. Insert the tassel skirt as described above. Cut a 16" (40 cm) length of binding thread, fold in two, and form a larkshead knot. Place the tassel on a table. Bind, working left to right (if left-handed, reverse the direction), as shown in figure 6.43:

- Slip the larkshead loop over the first bundle of skirt threads, about ½" (13 mm) from the tassel head. Pull the knot firmly around the group. Thread the loose ends of the binding thread through a tapestry needle

- Bring the threads over the second bundle. Stitch down between groups 2 and 3, coming up between groups 1 and 2 and splitting the two threads of the binding

- Stitch down between groups 3 and 4, coming up between groups 2 and 3, splitting the binding thread

- Continue this pattern (over two, under 1) until the last group has been covered. To join the binding, bring the threads over the first bundle and stitch down into the braid through the loop of the larkshead knot

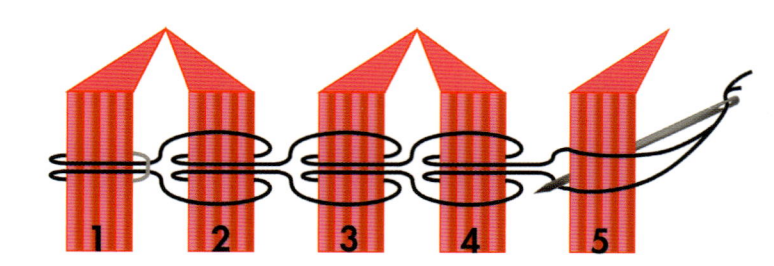

FIGURE 6.43.
Begin with a larkshead knot over bundle #1 and bind each bundle as shown.

## Working Braids over Sculptural Forms

Sculptural forms, like wooden pens, basket handles, and the lace bobbin seen in figure 6.44, can be inserted into braid structures that carry cores. Figure 6.45 shows a braiding card set up by Suzi Siorek for covering a form. (See Suzi's covered forms in *Andean Sling Braids, New Designs for Textile Artists*.[2]) The braid on the lace bobbin was made on the stand. A 16-strand design was set up for a blunt-end. The bobbin was inserted in the center of the warp and held in position with one hand, while the first few steps of the braid were worked with the other. The braid was continued beyond the bobbin in order to show the undistorted fret pattern. Twenty-four strand braids can also be used in this way.

**FIGURE 6.44.**
Lace bobbin covered with a 16-strand braid.

**FIGURE 6.45.**
Card set up for working a 24-strand braid around a wooden form. *Courtesy of Suzi Siorek.*

**FIGURE 6.46.**
Enlargement of a fragment of a Nasca mantle border of birds and flowers constructed with cross-knit looping. 2" × 19" (5 cm × 48.5 cm). *Courtesy of the British Museum (ref. Am1954, 05.512).*

## Cross-Knit Looping

A popular technique used during the Paracas and Nasca Periods for creating both simple and intricate edgings for mantles and other textiles was *cross-knit looping* (also referred to as needle knitting, loop stitch, knit stem-stitch, and *anillado cruzado* in Spanish):

> The needle knitting stitch itself is a variant of the cross-stitch . . . It may be employed to make a narrow line or a wide band, but it must be constructed over foundation material or a core. Specimens of needle knitting in the Early Nazca collection illustrate types of specially constructed foundations such as narrow tapes, shapes made by buttonhole stitchery, and twists of yarn for cores [some] required one, two, or three of these foundation types.[3]

The birds and plants in the Nasca mantle border (100 BCE–600 CE) shown in figure 6.46 measure 2" (5 cm) high. They were expertly stitched in cross-knit looping using a fine camelid yarn over a foundation of cotton.

### Using Tubular Cross-knit Looping to Finish or Embellish a Braid

Cross-knit looping can be worked in the round to enclose the cut ends of a braid where a tassel is not desired. It can also be used to decorate the head of a tassel. (See figure 6.47.)

The area to be stitched is first wrapped, usually with the same color as the stitching. In the first row, the needle is inserted into the wrapping. Subsequent rows are worked through the stitch above.

Thread a tapestry needle (size 18 for most braids) with the wrapping thread and anchor it in the braid about an inch from the end of the braid. Stitch into the braid, coming out close to the end. Firmly wrap the yarn around the braid for the desired length. Secure the end by stitching into the braid.

**Step 1.** Thread the needle with the yarn to be used for the cross-knit looping. Stitch right to left into the first corner of the braid just above the wrapping, leaving a 1" (2.5 cm) tail.

**Step 2.** Hold the tail to the left and bring the thread over the tail to form a loop and stitch into the next corner to the right. (See figure 6.48.)

**Step 3.** Leave the tail hanging down the length of the braid and stitch the remaining two corners.

**Step 4.** You will be back at the first stitch. Sew through the cross of the stitch above to form the next loop. (See figure 6.49.) Continue working anti-clockwise around the braid in this manner.

When the looping reaches the end of the wrapping, fasten off by stitching into the braid if this is a finish above a tassel. Or, to finish off a braid with a blunt end, work an additional row or two, pulling the loops to cover the end of the braid. Fasten off by stitching up into the braid. If stress will be put on the gathered stitches, stitch across the braid as well.

**FIGURE 6.47.**
Model of tubular cross-knit looping worked over a braid.

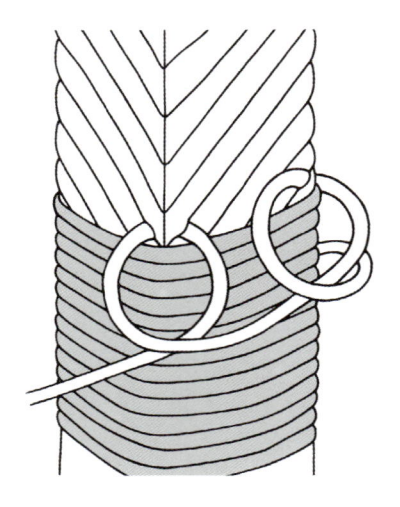

**FIGURE 6.48.**
Only the first row of cross-knit looping is stitched into the corners of the braid.

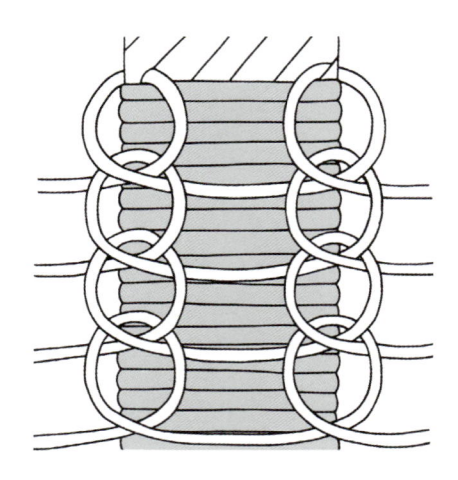

**FIGURE 6.49.**
Work proceeds in rounds with each loop stitched into the loop above, working from top to bottom.

**FIGURE 6.50.**
Sling cradle with cross-knit looping worked in colors that contrast with the woven edge.

## Finishes for Sling Cradles

In the Andean weaving tradition, it is customary to finish the selvages (outside edges) of cloth that will be subject to wear, such as shawls and carrying cloths. Similarly, the selvages of sling cradles are sometimes given an edge treatment. This may be more common in dance slings than in herding slings. Cross-knit looping, chain stitch, and applied braids are some of the techniques we have seen. Tassels are sometimes inserted through the edging as seen in the cradle in figure 6.50.

Andean sling makers choose edging colors and patterns to enhance the cradle's design. The colors of the cross-knit looping in figure 6.50 contrast with the weaving. This not only defines the edges of the cradle, it also emphasizes the positive/negative relationships of the shapes in the pattern. The cradle in figure 7.21 also uses the color of the stitching to enhance the cradle design. The cradle in figure 7.22 uses a different strategy. Here the colors for the 6-strand braid were chosen to coordinate with both the braid and the cradle designs.

### Cross-knit Looping Used for Finishing Sling Cradle Selvages

The cradle in figure 6.51 *(left)* has a cross-knit looped edging that was worked in multiple colors. Figure 6.51 *(right)* shows how the first loop and subsequent loops are stitched

**Step 1.** Thread the needle with the finishing yarn. With the edge of the cradle facing you, anchor the yarn by taking about a 1" (2.5 cm) stitch into the edge of the cradle, coming out at the right-hand side at "start." Leave a 1" tail that will be clipped later.

**Step 2.** Stitch from right-to-left coming out at #1.

**Step 3.** Stitch into the edge of the cradle from right-to-left at #2 to form the top of the initial loop.

**Step 4.** Insert the needle at #3 making a diagonal stitch coming out at #4.

**Step 5.** Take the needle *under* the first loop from right to left; then make the next diagonal stitch to finish one loop and begin the next.

Continue repeating Step 5. (The first loop may not stay as open and neat as subsequent loops.) The color can be changed by ending color A on the right, anchoring color B in the cradle, and bringing it up on the left. Then continue stitching. If color A is to be reintroduced, it can be carried on the edge of the cradle under the stitching or it can be stitched into the cradle and brought up where needed.

The cradle in figure 6.52 *(left)* is finished with a simple 6-strand herringbone braid that has been sewn on. The herringbone design is a very early braid common in the Paracas burials. (See figure 1.27, Chapter 1, *A History of the Sling.*) In this example, two groups of three colors are used. Figure 6.52 *(right)* shows the outermost thread on the right, crossing to the inside position on the left. Next, the outermost thread on the left will cross to the right. These two steps are repeated. This braid can be made with more strands than our example.

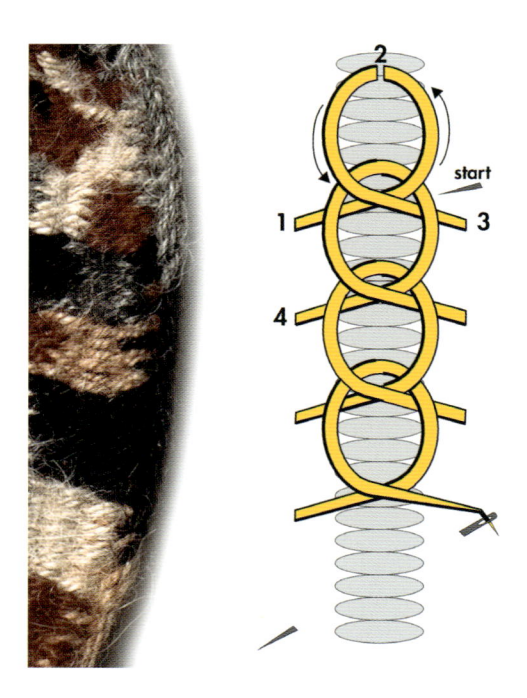

FIGURE 6.51.
*(Left)* Cross-knit looping stitched into the selvages of a tapestry woven cradle. *(Right)* Diagram for stitching cross-knit looping stitched onto the edge of a cradle.

FIGURE 6.52.
*(Left)* A 6-strand braid is sewn onto the edge of a tapestry woven cradle. *(Right)* The herringbone braid is worked from the outside to the center.

# 7. MAKING AN ANDEAN-STYLE SLING

**FIGURE 7.1.**
An experienced female slinger demonstrates the distance a stone can be slung in Rio Ramis drainage area, Department of Puno, Peru. *Courtesy of Margaret Brown-Vega.*

A sling is a weapon, herding tool, or ceremonial object. Whether simply or skillfully made, the sling should feel balanced and work as an extension of the slinger's arm. Most slings consist of two lengths of cord, on either side of a wider portion called a cradle or pouch. (See figure 7.3.)

The cradle is the focal point of a sling. Its function is to hold a missile securely until it is released and sent to its target. Dance slings may have cradles that are more decorative and less functional, as seen in the Bolivian cradle shown in figure 1.39. Andean slings, like the one used in figure 7.1, feature a finger loop that keeps the sling in the slinger's hand once the missile is released.

Master weaver Oscar Huarancca Gutierrez of Ayacucho, Peru, in a recent conversation with the authors, explained that among Andean sling makers there are regional differences in sling design, as well as individual preferences in braid pattern, color choice, and cradle design. Likewise, each slinger develops their own style for slinging stones. He made it clear that the sling is still very much respected for its potential as a powerful weapon.

## Traditional Macusani and Ilave Sling Making Methods

In her 1982 article "Sling Braids in the Macusani Area of Peru," Elayne Zorn describes two sling styles and their differing methods of construction.[1] One major difference is the thickness of the sling cords. The slings she was taught to make in Macusani, featured one thin cord and one highly patterned thick cord. The slings she described that came from a community in Ilave, near Lake Titicaca, featured cords of the same size that increased in diameter as they neared the cradle.

Ilave-style slings either begin with a braided loop or incorporate a split loop within a few inches of the cord's beginning. Macusani-style slings do not customarily start with a braided loop but often have a split loop. Both styles have patterned cradles woven in *weft-faced, plain weave tapestry*.[2] (See *Tapestry Weaving Basics*.) Most cradles incorporate a slit to allow irregularly shaped stones to sit securely. Edges of the pouches of both styles are sometimes reinforced with stitching or with a narrow braid that has been sewn on.[3]

**FIGURE 7.2.**
Macusani-style sling, worked from the center out to the tasseled ends, features thick and thin cords, 5 ft. (1.52 m) long including a 6" tassel. *Collection of R. Owen.*

For the Macusani-style sling, two separate warps are prepared: warp A, for a thicker, more elaborately patterned braid made of 32 strands or more; and warp B, for the cradle and a thinner braid of 5, 7, 8, or 16 strands. (See figure 7.2.) The two warps are folded in half and linked together at the center. Warp B is folded and wrapped to create a temporary handle while warp A is braided from the cradle to the end of the first sling cord. A finger loop is made by splitting the braid into two smaller braids, working them for a short distance, and rejoining them near the end of the cord.[4] The work is then turned around and warp B is set up for weaving of the cradle. Its width is gradually increased and then decreased, ending with a length of thinner braid.[5] A more detailed explanation of making this style sling is described by Laverne Waddington later in the chapter.

Ilave-style slings are worked from one end to the other. They have one base warp that is used throughout the length of the sling and several others that are added to increase the diameter, or to add additional colors or patterns to the braid. Our model is based on the Ilave-style sling shown in figure 7.3. In this instance, warp A is used throughout and is measured twice the finished length. The shorter warps, B, C, and D, are used to increase the diameter of the braid, add a color, and to provide extra warp for the cradle. *The number of warp groups that are added before the cradle and what is done to remove extra threads when they are not needed before or after the cradle is woven varies from sling to sling.*

Braiding begins in the center of warp A with a 4" (10 cm) length of 8-strand braid. The braid is folded in half to form the finger loop. Then a length of 16-strand braid is worked with the combined ends. Warps B and C are added to increase the diameter of the cord with short lengths of core carrying 24-strand patterns. The braid warps are then re-grouped to form the cradle's warp, e.g., 8 groups of 4 warps each. On completion of the cradle, the warp is gathered, warp D is added, and the braid is continued, gradually decreasing the number of threads back to 16 strands. The second cord is braided to match the first cord. Near the end, the 16 strands may be divided into two 8-strand braids and rejoined to make a split loop.

The sample sling shown in figure 7.4a uses 8-, 16-, and 24-strand patterns in the cords, and dovetail joins in the tapestry-woven cradle. In this sling, four red threads are cut from the warp before the cradle is woven. The remaining 32 warp threads are divided to make the eight cradle warps. (Instead of cutting extra warps out before weaving the cradle, they could be added to the warp groups at the selvages. We were unable to determine if there is a tradition of cutting extra warps versus distributing the extra threads in the cradle warp.)

The instructions for making the Ilave-style sling shown in figure 7.4a are written for both the core frame and the card. Braid Designs 8.2.2, 16.3.1, and 24.12.2 were chosen for the sling cords and Cradle Design 2 for the cradle. When other braid designs are substituted, the way in which threads are added and/or decreased along the length of braid may need to be adjusted. Figure 7.4b shows a sling made on a card using Designs 8.1.1, 16.3.4, 24.12.5, and 24.12.7, with Cradle Design 1. The yarn was an English rug wool.

Sling 7.4a was made with a 3-ply rug yarn (80% wool, 20% nylon) in rusty red, dark brown, and white. For core frame braiders, the bobbin weight was 70 grams and the counterbalance weight ratio was reduced to make a firmer structure. The changes in the counterbalance weights used in each section of the sling are noted in the text.

#### FIGURE 7.3.
An Ilave-style sling starts with an 8-strand finger loop followed by 16-, 24-, and 36- strand braids. The cradle was woven over eight warp groups with a slit tapestry technique.
*Collection of R. Owen.*

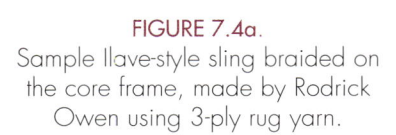

#### FIGURE 7.4a.
Sample Ilave-style sling braided on the core frame, made by Rodrick Owen using 3-ply rug yarn.

**FIGURE 7.4b.**
Ilave-style sling braided on a card using 100% wool rug yarn, by Terry Flynn.

## Calculating the Warp Length for a Sample Sling

Braiders in the Andes prepare their warps with and without the aid of warping posts. Horst Natchitgall described a traditional method of warping for a sling used in Salinas (near Arequipa, Peru) as follows: "To prepare the yarns for braiding, they are measured by arm spreads: for a smaller sling, four lengths, each of a different color and each of 8 arm-spreads; for a longer sling, four lengths each of 12 arm-spreads."[6] On average, eight arm-spreads equal 12 yards (11 m). The Andean herding slings in the author's collection vary considerably in length from 52–80" (132–203 cm). The sample sling shown in figure 7.4a has a finished length of 5 ft. (1.5 m). The longest sections of its warp were 18 ft. (5.5 m).

### Prepare Four Warp Sections

**Warp A.** Cut four red and four brown warps 18 ft. (5.5 m) each. This warp is used throughout the entire length of the sling to make the 8-strand loop, cord #1, the cradle, and cord #2.

**Warp B.** Cut two red and two brown warps 10 ft. (3 m) each. This warp is shorter and will be used to increase the 16-strand braid to 24 strands.

**Warp C.** Cut six white warps 7 ft. (2.10 m). This warp will be added to make the brown and white 24-strand braid on either side of the cradle.

**Warp D.** Cut two red warps 7 ft. (2.2 m) long. This warp is used to replace threads that will be removed before the cradle is woven.

# Making the Finger Loop and First Sling Cord

Prepare warp A as described in Chapter 6, *Preparing the warp and setting up for a loop*, referring to figure 6.4. (Stand braiders will attach the counterbalance bag to the warp with 10 oz. [290 g] of counterbalance weight). All braiders make approximately 4" (10 cm) of Design 8.2.2 in figure 6.22. (To gauge the correct size for the loop, the folded braid should have room for two fingers to be inserted.)

## Making the 8-strand Finger Loop

To prevent the braid from unraveling when the threads are being rearranged for joining the loop, use a separate thread to tie it off on the underside of the stand or card. Bring the loose warp up through the center hole. (Stand braiders will reposition the knitting needle as shown in figure 6.10.) To interlace the two ends of the braid neatly, identify the lower threads and arrange them according to the set up for Design 16.3.1. (See figure 6.20.)

Braid the first four steps without a counterbalance, carefully snug up the threads as necessary and braid four more steps. (Attach the counterbalance bag with an S-hook loop and increase the weight to 16 oz. [450 g]). Continue Design 16.3.1 for 19" (48 cm) from the loop, or for the desired length, ending with Step 4.

## Transitioning from 16 Strands to 24 Strands

To increase the braid from 16 to 24 strands, warp B (two red and two brown warps) will be added. (Stand braiders will attach bobbins to both ends of each warp as described in Chapter 6, *Starting a 16-strand Braid with a Blunt End, Method B*).

Follow Steps 1–12 in figure 7.5 to add warp B. At Step 1 place the brown threads on the stand or card, followed by the reds and work Step 1a to hold them in place. Then at Step 2 place the brown, then the red threads, and work Step 2a. Work Steps 3–12. (Leave the counterbalance weight unchanged at 16 oz. [450 g].)

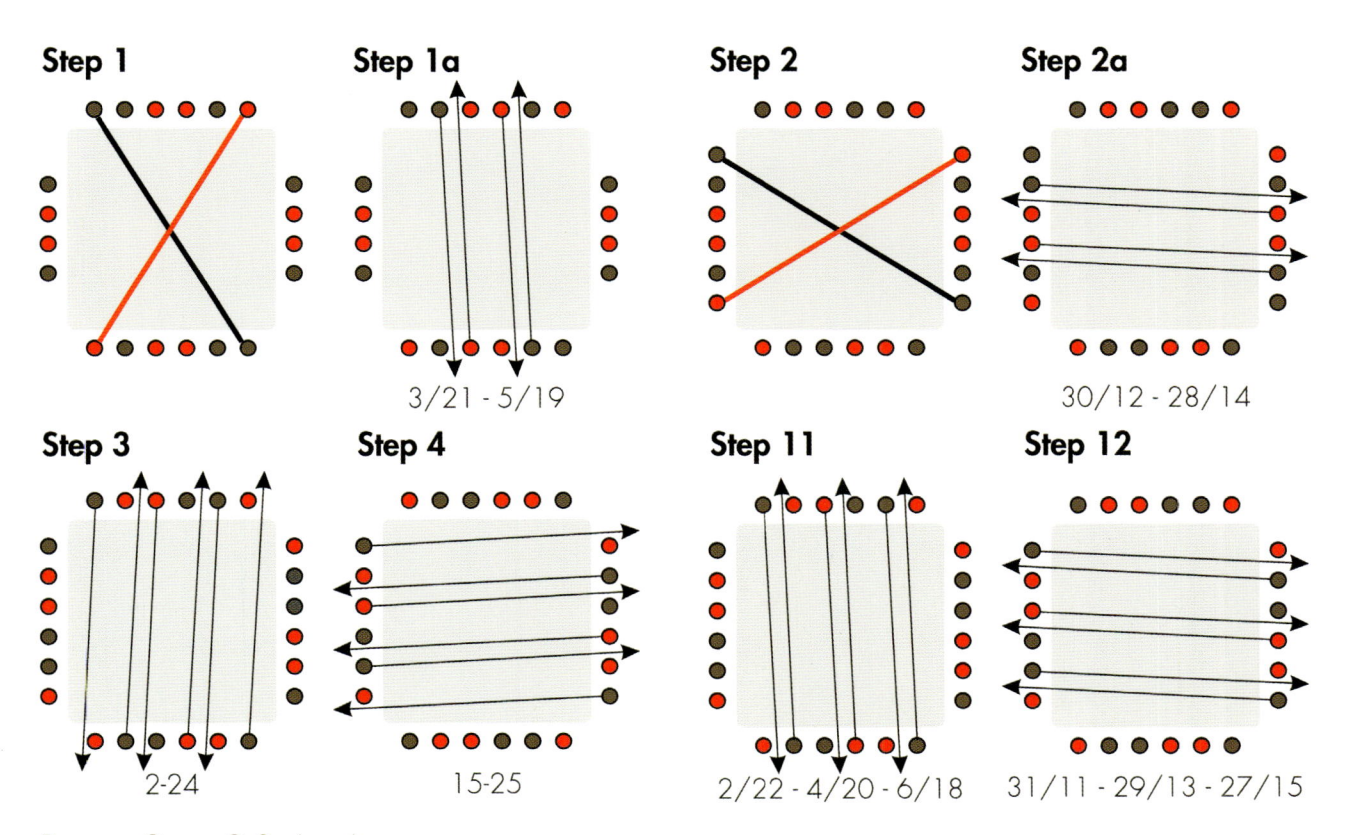

FIGURE 7.5.
Red and brown threads added to make a 24-strand braid.

## Adding 12 White Threads

Figure 7.6 shows the steps for adding white threads to the structure. Each white strand is centered and will replace an outside or inside pair of red threads. The red threads will remain in the core until the threads are regrouped to create the warp for the cradle. This process has two parts. In Steps 13–18 the new pair is inserted and the displaced threads move to the core. Steps 13a–18a are partial steps used to braid the remaining threads.

**Stand braiders.**

**Steps 13–18.** Attach a bobbin to each end of the six white warps. The white threads are placed across the braiding stand (as indicated by the pair of diagonal lines) and the red threads are lifted onto the harness. Note: When adding the white threads for Steps 15–18, thread them through the suspended red harness threads.

**Steps 13a–18a.** After inserting each white thread, work the corresponding step to secure the new pairs in place. The counterbalance remains at 16 oz. (450 g) until Step 22 is completed; then increase the weight to 49 oz. (1400 g).

**Card braiders.**

**Step 13.** Find the middle of the white thread and place it diagonally across the card in slots 1 and 17. Bring the red threads from slots 7 and 23 to the core.

**Step 13a.** Flip the core to the left and partially work moves 7-17, skipping the first and last moves: 6-7, 19-6, 5-19, flip core, 4-5, 21-4, 3-21, 2-3, reposition threads.

**Step 14.** Insert at white thread in slots 16 and 32. Bring the red threads from slots 10 and 26 to the core.

**Step 14a.** Work a partial 10-32: 11-10, 30-11, 12-30, flip core, 13-12, 28-13, 14-28, 15-14, 26-15. Reposition threads.

**Step 15.** Part the core and lay a white thread in the center.

**Step 15a.** Work a partial 7-17 as follows: 7-17, 6-7, 19-6, 5 to the core, white thread in 19, 4-5, white thread in 4, 21 to the core, 3-21, 2-3, 23-2. Reposition threads.

**Step 16.** Part the core and lay a white thread in the center. Bring red threads 12 and 28 to the core.

**Step 16a.** Work partial 10-32: flip core, 10-32, 11-10, 30-11, 12-13, white thread in 13, white thread in 30, flip core, 14-28, 15-14, 26-15. Reposition threads.

Insert the last two threads in Steps 17 and 18 by repeating Steps 13–14 and Steps 13a–14b, parting the core when inserting the threads.

**All braiders.**

Finish by working Steps 19–22.

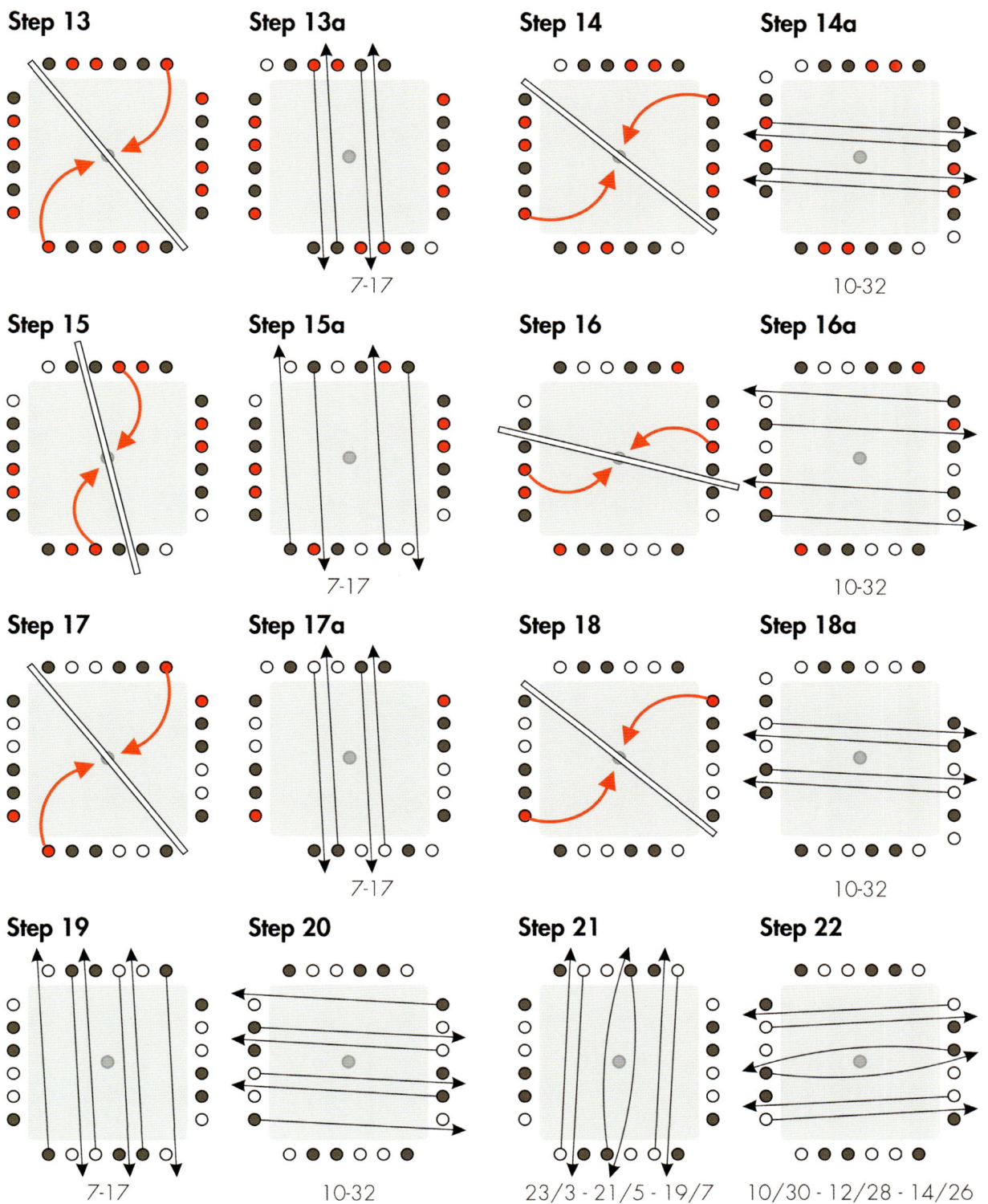

FIGURE 7.6.
Steps for adding 12 white threads to the braid and moving 12 red threads to the core.

163

Braiding the 24-strand Diamond Pattern

Make three white and brown diamond patterns before beginning the cradle. Steps 23–42, shown in figure 7.7, are repeated twice to make the first two diamonds, thus ending on Step 62. The third diamond is made with Steps 63–79, shown in figure 7.8.

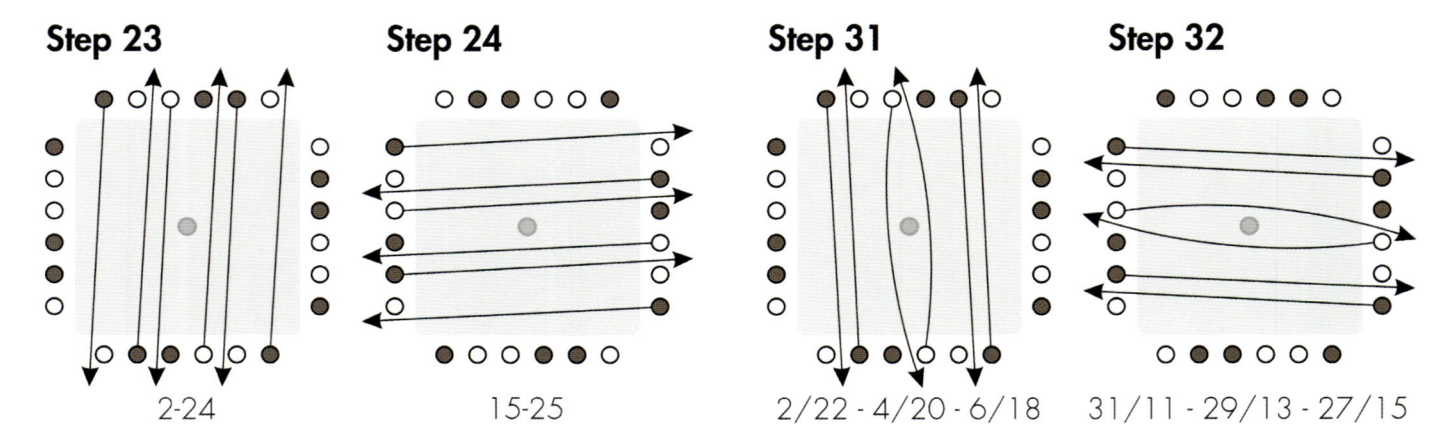

**Step 23**

2-24

**Step 24**

15-25

**Step 31**

2/22 - 4/20 - 6/18

**Step 32**

31/11 - 29/13 - 27/15

Repeat Steps 23 and 24 x 4, and then work Steps 31 and 32.

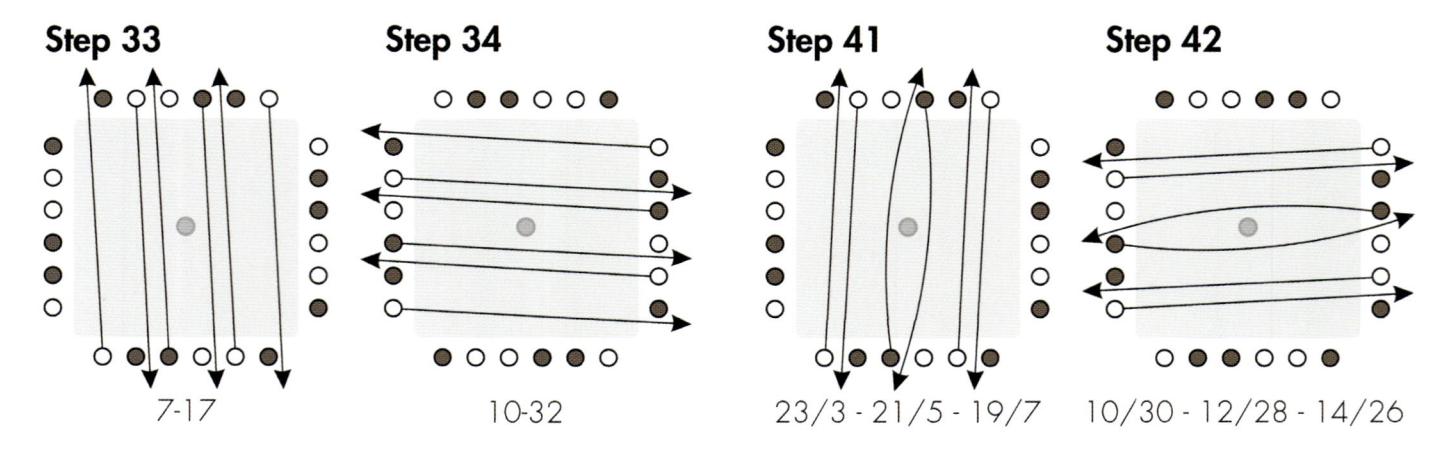

**Step 33**

7-17

**Step 34**

10-32

**Step 41**

23/3 - 21/5 - 19/7

**Step 42**

10/30 - 12/28 - 14/26

Repeat Steps 33 and 34 x 4, and then work Steps 41 and 42.

FIGURE 7.7.
Make the first two brown and white diamonds by repeating Steps 23-42 twice.

## Preparing the Warp for Weaving the Cradle

To ensure a smooth transition between the last braiding move and the first wrapping of the cradle, always end the braiding sequence with a N/S step as seen in Step 79 (figure 7.8). Note: Finishing with an E/W step or an inversion step would create crossed threads, preventing a smooth transition from the braid to the cradle.

The core threads must now be rearranged to provide eight warp groups for the cradle. Refer to figures 7.9, 7.10, and 7.10a. For both methods, eight core threads are added to the regrouped threads on the stand or card and the remaining four core threads are cut carefully near the point of braiding. These are replaced after the cradle is woven. (Alternately they can be brought down to join the threads that will be on the selvages of the cradle making these warp groups slightly thicker.)

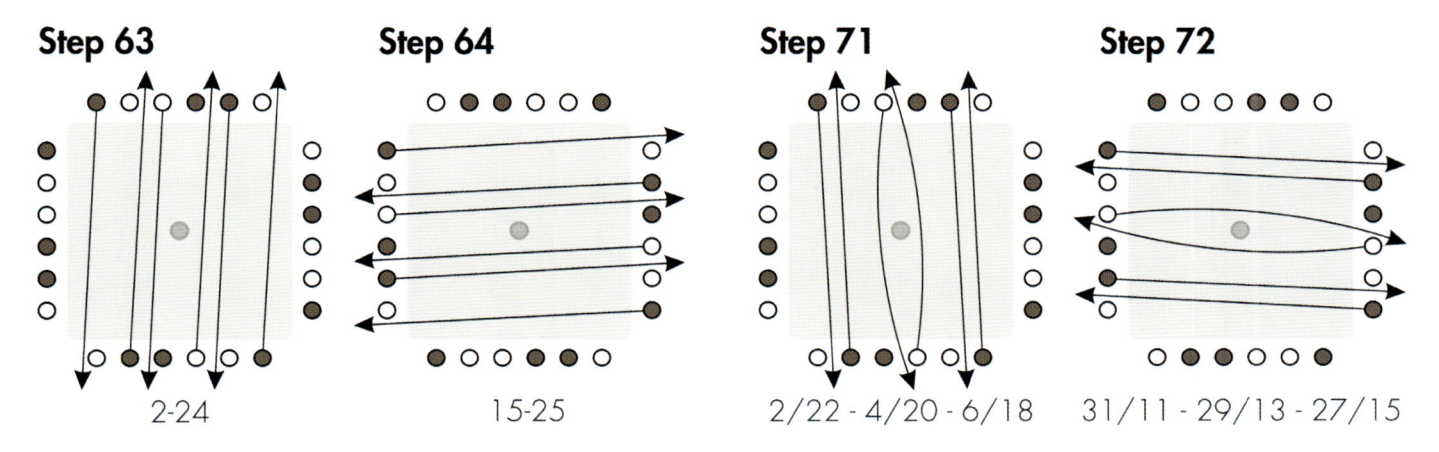

**Step 63**  **Step 64**  **Step 71**  **Step 72**

2-24   15-25   2/22 - 4/20 - 6/18   31/11 - 29/13 - 27/15

Repeat Steps 63 and 64 x 4, and then work Steps 71 and 72

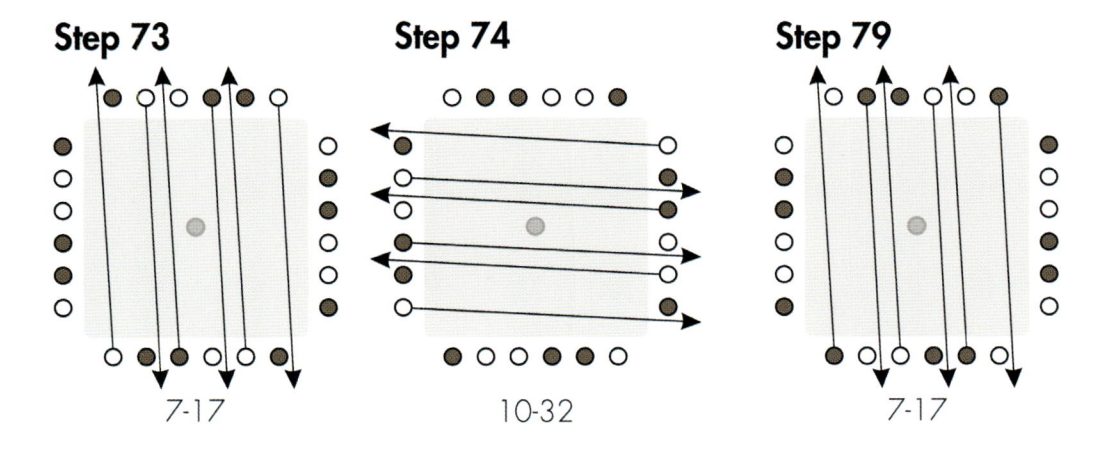

**Step 73**  **Step 74**  **Step 79**

7-17   10-32   7-17

Repeat Steps 73 and 74 x 3, and then work Step 79

FIGURE 7.8.
To make the last diamond before the cradle is started, work Steps 63–79.

**Stand braiders.** Bring down eight red core threads (from the two outermost threads on each arm of the harness). Place four in the center of the N threads and four in the center of the S threads. Move the E/W threads to join the N and S threads as indicated by the arrows in figure 7.10. Carefully cut the remaining four core threads near the point of braiding. Lower all the bobbins to the base of the stand, *adjusting the height of the counterbalance* to keep the threads under tension.

**Card braiders.** Rearrange the N/S threads to make room for the core threads. Move:

- N/W threads 4-3, 31-2, 30 and 29-1

- N/E threads 5-6, 10-7, 11 and 12-8

- S/W threads 21-22, 26-23, 27 and 28-24

- S/E threads 20-19, 15-18, 13 and 14-17

- Place two core threads in 4, 5, 21, and 20. Cut the remaining core threads at the point of braiding.

**FIGURE 7.10.**
Core threads and E/W threads are moved to the N and S of the stand.

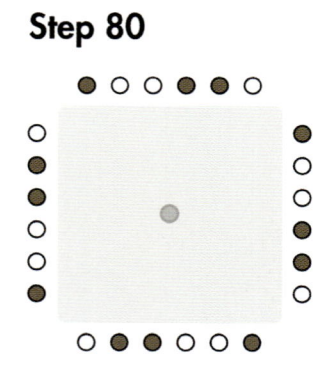

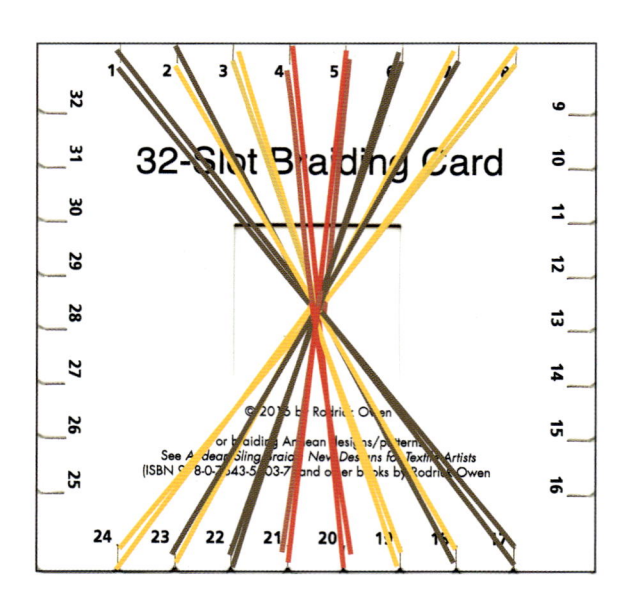

**FIGURE 7.9.**
Thread arrangement after completion of Step 79.

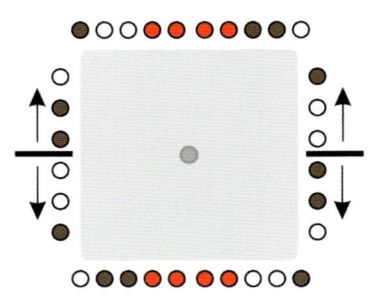

**FIGURE 7.10a.**
Card set up with two threads in each slot.

## Binding the Thread Groups

Next, pairs of N threads will be brought to join the S threads, forming four-thread groups that will serve as the warps for the cradle. Figure 7.11 shows the color order in which each group of four threads are to be combined. Loosely wrap each group and tie it off with a surgeon's knot, working from the center out. This keeps the threads in order before twining. (See figure 7.12.)

- Bring the #1 pair of red threads to join the pair in the S

- Tie a loose binding around the group as shown in figure 7.12. Position the binding along the outer edge of the stand or card. (Card braiders leave a few threads in the slot to keep the bundle in position.)

- Continue binding each group in the order shown in figure 7.11

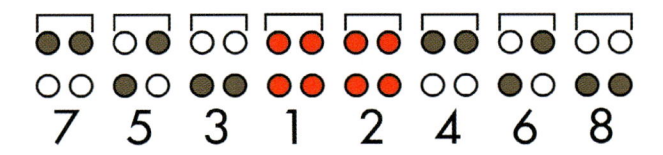

**FIGURE 7.11.**
Order for arranging threads for the cradle's warp.

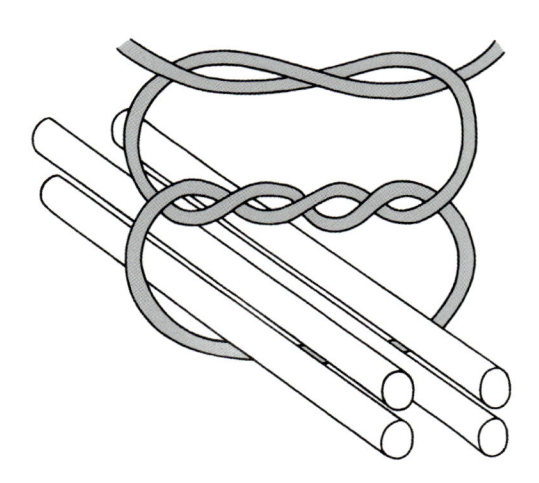

**FIGURE 7.12.**
Loosely wrap each four-thread group and secure with a surgeon's knot.

## Organizing Warp Groups by Twining

Cut two 12" (30 cm) pieces of binding thread. Fold the pair over the #7 warp group. Then counter twine around the warp by bringing the bottom pair up between the top pair encasing each warp group. (See figure 7.13.) Adjust the tightness of the twining so that it can move freely up and down the warp. Slide the twining down about 3" (7 cm) below the edge of the stand or card. Tie the loose ends in a square (reef) knot. (Stand braiders remove the bobbins and release the counterbalance.) All braiders remove the warp from the stand or card.

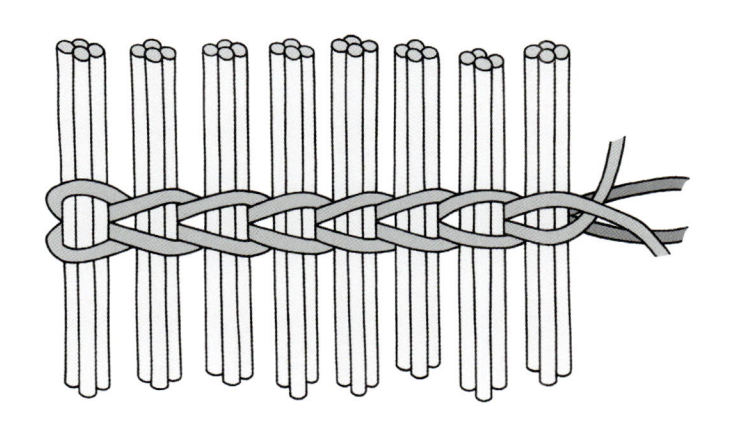

**FIGURE 7.13.**
Counter twine four-thread groups of warp before releasing the bobbins.

## Weaving the Cradle

The cradle of the sling will be woven. While the structure of the braided cord gives the sling some ability to stretch and bounce back, tapestry technique makes the cradle firm and abrasion resistant.

### Tapestry Weaving Basics

The cradles of Andean slings are woven in the tapestry technique called *weft-faced plain weave*. Figure 7.14 shows the *weft* (yarn that interlaces with the warp) traveling over one warp and under the next, forming the plain weave structure. (Each vertical thread in the diagram represents a warp group.) Unlike the weft shown in the diagram, tapestry wefts must be compacted so that they completely cover the warp.

The simplest patterns are made by weaving stripes from one selvage (edge) to the other. More than one color weft can be used in a row to create designs, usually geometric in Andean slings. These designs range from simple to complex. The number of color changes, called *joins*, determines the level of difficulty of the weaving.

The following methods for changing color are often used in Andean slings. In *slit tapestry* technique (also called *discontinuous weft* or *kilim*), each color turns back to weave across its warps, leaving a slit between the color areas. (See figure 7.15.) In *dovetail* technique, at the end of a row, the weft crosses over or under its neighbor's end warp before turning back to weave its color area. (See figure 7.16.) In *interlocking* technique, the wefts meet and turn around each other before going back to weave across their warp groups. (See figure 7.17.)

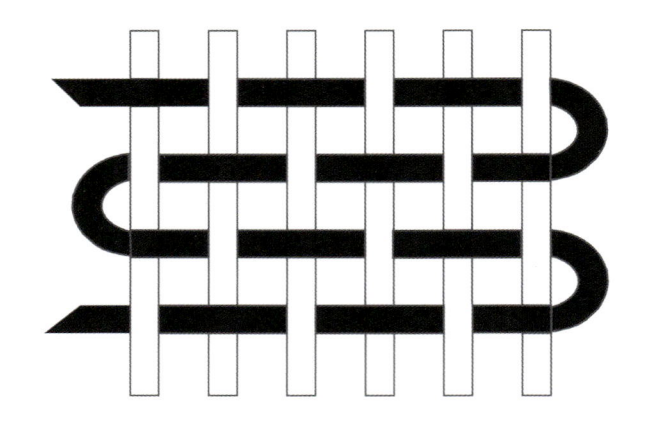

**FIGURE 7.14.**
Balanced plain weave. For tapestry, the weft is beaten down to completely cover the warp.

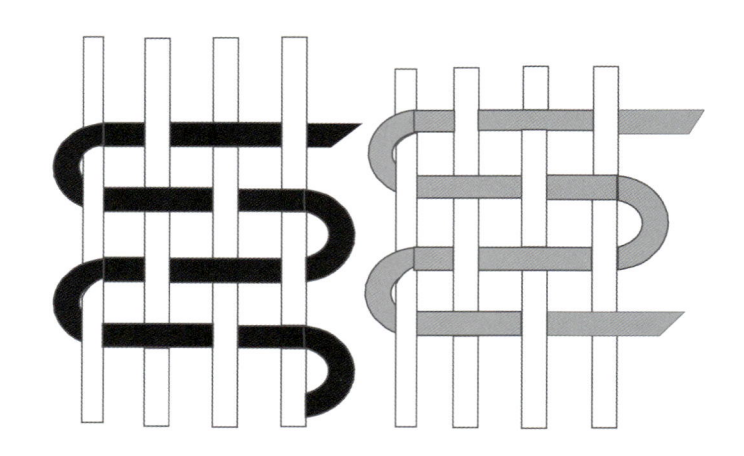

**FIGURE 7.15.**
To weave a slit in the cradle, two wefts weave independently.

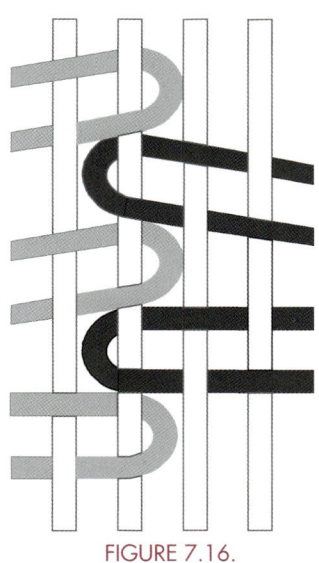

**FIGURE 7.16.**
Dovetailing is used to prevent slits from forming between two color areas.

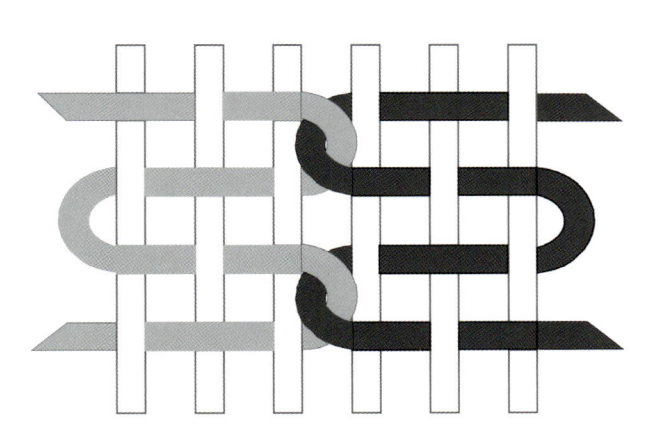

**FIGURE 7.17.**
Interlocking joins are used to prevent slits from forming.

## Cradle Design

The instructions are written for two basic cradle designs. Cradle Design 1 has a bi-colored cradle and is similar to the Andean cradle shown in figure 7.18. Cradle Design 2 features a central diamond pattern with a short slit in the pouch, like the one shown in figure 7.19.

The first step in designing a cradle is to decide on the cradle's length and width. This is influenced by the size of the missile and the slinger's preference. Cradle measurements for the Ilave-style herding slings in the Owen collection vary considerably. Refer to the cradle gallery to compare sizes; additional information on each sling is given in Appendix F.

Our examples measure as follows:

Cradle 1. 6½" × 2" (16.5 cm × 5 cm)

Cradle 2. 5½" × 2¼" (14 cm × 6 cm)

**FIGURE 7.18.**
Cradle Design 1 is seen in both Macusani- and Ilave-style slings. *Collection of R. Owen.*

**FIGURE 7.19.**
Cradle Design 2, woven in slit tapestry technique, features a central diamond bordered by stripes with a short slit.

Andean sling cradles are typically woven over 8 warps, sometimes 12 or 16. We recommend starting with a design for eight warps and using the same yarn as is used for the warp. (For more complex patterns, a thinner weft would be used.) If a slit is desired, decide on the length before the tapestry design is chosen since the slit will affect the pattern placement. The slit lengths vary according to the size of the missile.

Andean sling makers do not plot their designs on paper as this is obviously not part of their textile tradition. The shape and size of the patterns are monitored and adjusted as the weaving progresses to the center. Then the design is reversed. Western sling makers may want to sketch the cradle design on graph paper as would be done for tapestry. Use a set number of squares to represent each warp group. It is helpful to draw this to scale.

For your first cradle, we suggest you follow our examples, substituting colors of your choice, or use stripes and simple geometric patterns. Once the design is determined, the warp is set up for weaving. Mark the halfway point of the cradle on the center warp with a permanent marker.

## Cradle Gallery

**FIGURE 7.20.**
Stepped diamond pattern worked over eight warps. The cradle measures 4½" (11.5 cm) long × 1½" (4 cm) wide between the braided cords. (See figure 1.40 for the full sling.)

**FIGURE 7.21.**
Cradle with a short slit and a pattern of squares woven with dovetail joins. The pouch measures 4" (10 cm) long × 1½" (4 cm) wide between the braided cords. (Also see figure 7.2.)

**FIGURE 7.22.**
Cradle with outlined diamonds woven with a long slit. The pouch measures 6" (15 cm) long × 2½" (6.5 cm) wide between the braided cords. (See Chapter 1, figure 1.44.)

**FIGURE 7.23.**
Fine threads used for weft in a cradle with outlined diamonds and wedges. The cradle measures 6½" (16.5 cm) long × 1½" (4 cm) wide between the braided cords. (See Chapter 1, figure 1.43.)

## Widening the Warp Before Weaving the Pattern

The procedure for beginning the wrapping and weaving is the same for both Cradle Designs 1 and 2. The braid is wrapped for a short distance to prevent the braid's stitches from loosening while the warp is set up for weaving the cradle. The first stages of weaving gradually spread the warp by dividing it into two, four, and then eight warp groups. (See figure 7.24.)

In order to completely cover the warp, use a small tapestry beater, kitchen fork, or wide-toothed comb to beat in the weft. Another method is to use a weaver's shed stick to beat it in. (See figure 7.26.1.) In tapestry weaving, the warp yarns stay straight and the weft yarns must be given enough ease to bend around them. This is done by angling the yarn as seen in figures 7.24 and 7.26.1. Experiment to find an angle that allows the weft to pack down easily while keeping the selvages neat. It may take a little practice to do both.

In order to avoid constantly cutting and rejoining threads, colors that are not in use can be carried alongside warp threads. Figure 7.24 shows the inactive yellow and red threads being carried alongside warp #1. They can be easily brought back into the weaving when needed.

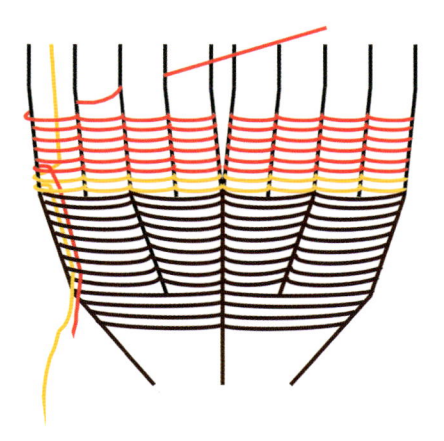

**FIGURE 7.24.**
New wefts can be added by stitching into the weaving, leaving a tail that is trimmed later. When that weft is no longer needed, it can run alongside a warp thread until required.

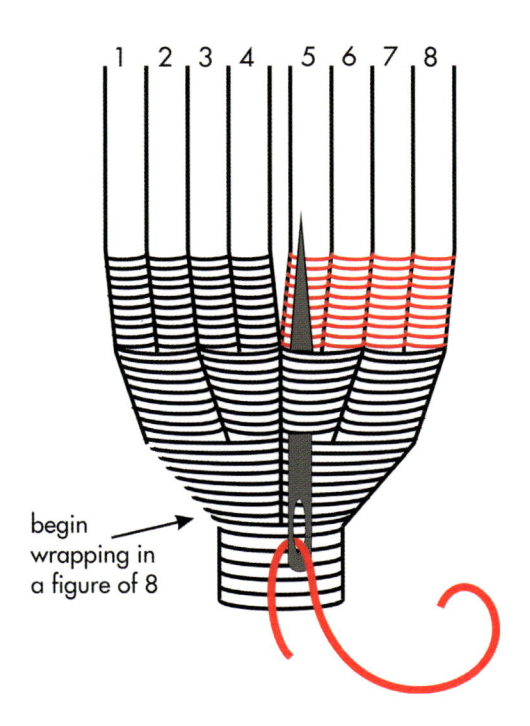

**FIGURE 7.26.**
Cradle Design 2 features a central diamond pattern divided by a slit.

**FIGURE 7.25.**
Cradle Design 1 features a bi-colored cradle. Both cradles are started by wrapping around the end of the braid, dividing the warp into two sections that are wrapped in a figure-of-eight, and then further divided until there are eight warp groups.

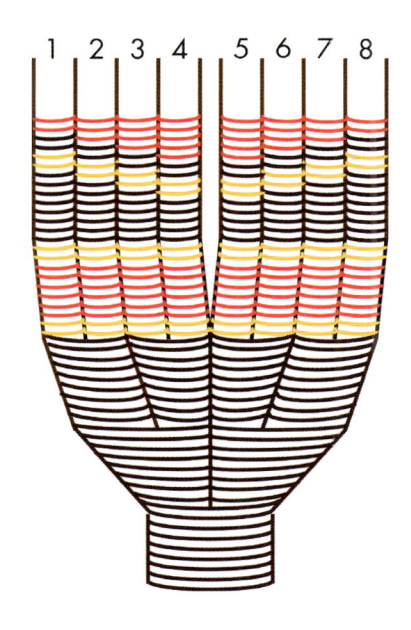

The cradle diagrams were designed for the 3-ply rug yarn that was used in our sample slings. (See figure 7.19.) The number of rows shown in the diagrams corresponds to this size yarn. *For Design 2, if a thinner yarn is used, you will need to increase the number of rows in some areas to avoid having a shortened cradle with flattened shapes.*

Cut a 1 yard (1 m) length of yarn. Thread a tapestry needle and stitch into the braid about an inch away from the unbraided warp, bringing the needle out ¼" from the end of the braid. (Leave a tail; it can be trimmed later.) To stabilize the braid as you wrap, you can anchor the braid between your knees while holding the loose warp in one hand. Remove the needle and tightly wrap the braid for about 5/8" (1.5 cm). Tie two half hitches (or a knot that can easily be untied) in order to keep tension on the wrap while the warp is set up for weaving.

## Setting Up the Warp for Weaving

In order to keep both hands free to manipulate the cradle's weft, the warp is stabilized in a way similar to that used for backstrap weaving. The unbraided warp must be anchored to something so that you can keep tension on it by adjusting your body's position as you weave. The weft is beaten in toward the body as the cradle is woven.

To gauge the distance needed for the warp tether, tuck the loop end of the braid into your belt leaving about 5–7" (13–18 cm) of the cradle end of the braid in front of you. Position the tether roughly 24" (60 cm) from your waist.

In the Andes, the warp may be stretched out horizontally by dividing the warp and tying it to two stakes pounded into the ground.[6] Western braiders can separate the warp into two groups and tie them to the back of a chair or to clamps or warping posts secured to a table. (See figure 7.26.1.) Unless you have backstrap weaving experience, expect that you will have to make some adjustments to your setup in order to find a position from which you can comfortably control the warp and easily see the area that is being woven. (If the warp tension becomes uneven during weaving, untie the warp, comb it out, and re-tether it.)

**FIGURE 7.26.1.**
Warp set up for weaving the cradle.

**Step 1.** Bring the wrapping yarn up between the 4th and 5th warp groups. (See figure 7.25.) Wind firmly around the two groups in a figure-of-eight until about ½" (1.3 cm) has been bound. Line up the weft threads, one after the other without overlapping, ending at the left selvage.

**Step 2.** Re-divide the warps into four groups as follows: 1/2, 3/4, 5/6 and 7/8. Weave over and under in plain weave for the desired number of rows.

**Step 3.** Divide the warp into the final eight groups and continue in plain weave. (See figure 7.14.)

## Weaving the Pattern and Creating the Slit

Begin the patterned section of the cradle following Design 1 or 2. (See figures 7.25 and 7.26.)

*Cradle design 1.*

**Step 1.** Starting at the left selvage, weave across warps 1–4; put black to the side.

**Step 2.** Thread a tapestry needle with red. Stitch into the cradle wrapping, alongside the warp, bringing the needle out near warp #5 as seen in figure 7.25. (Leave a tail that can be trimmed later.) Weave across warps 5–8, continuing the plain weave pattern, i.e., if the weft traveled under the even number threads on the left side, they must do the same on the right. (This will avoid an error in the pattern as seen in figure 7.18.)

**Step 3.** Each color area is now woven over its own set of warps. Work a section on the left, then work the right to match.

**Step 4.** When the slit is the desired length, go to *Closing the Slit and Tapering the Cradle*.

*Cradle design 2.*

(Refer to figure 7.26.) The initial weaving is identical to that of Cradle 1. After the slit is established, a stripe or shape is woven on the left and then mirrored on the right, working back and forth until the pattern is complete and the two sides are ready to rejoin.

## Dividing for the Slit and Weaving the Triangles

Leave the brown weft at the left selvage. Anchor a strand of yellow by stitching through the cradle and come up at warp #1. (See figure 7.24.)

**Step 1.** Weave across warps #1–4. Weave back to the left selvage. Repeat these two rows to create one narrow stripe that appears as a thin double line on the front and back of the sling. (Enclose the inactive brown thread as described above.) Begin the right side of the slit by bringing up a separate strand of yellow at warp #5. Weave the right side to match the left.

**Step 2.** On the left, drop the yellow and anchor a strand of red. Weave a wider stripe of 6 rows. (Carry inactive brown and yellow wefts with warp #1.) *Continue matching the left-side pattern on the right.*

**Step 3.** Swap the red for yellow and weave 4 rows.

**Step 4.** Swap yellow for brown and weave 6 rows to create the base of the triangle.

**Step 5.** Build the first step of the half diamond by weaving across warps #1–3, turning and weaving back. Repeat twice. Mirror this on the right by weaving across warps #8–6.

**Step 6.** Continue weaving around one less warp per step, matching the height of previous steps, until one warp remains. Wrap this warp to match in height.

## Outlining and Filling the Diamond

Starting with yellow at the selvage, wrap around each warp twice. (See figure 7.26.) Adjust the direction of the wrap to prevent floats on the reverse side of the cradle, i.e., wrap clockwise around one warp and anticlockwise around its neighbor. (Figure 7.27 shows outlining in progress on a different cradle.) Switch to brown and weave another outline. To fill in the half diamond, bring up the red at warp #4.

**Step 1.** Weave in a figure-of-eight around warps #3 and #4, filling in the gap and ending at warp #4.

**Step 2.** Weave across warps #4–#2, turning around warp #2, and returning to the slit. Repeat to fill in the space.

**Step 3.** Weave across warps #4–#1 and repeat to weave the center of the half diamond.

**Step 4.** To mirror the design and weave the second half of the cradle up to where the slit is to be closed, reverse the shaping process by weaving progressively shorter rows of red. Outline with brown and then yellow. Continue reversing the pattern until it is time to join the two sides.

**Step 5.** Weave the right to match the left.

## Closing the Slit and Tapering the Cradle

When the slit is the desired length, leave one weft at a selvage. Before ending the second weft, check to see that the upper and lower warps alternate in plain weave order from one selvage to the other. If they do not, an error will occur on one side of the cradle. This appears as small gaps as seen in the left-hand side of the slit in Cradle #1. (See figure 7.18.) If a correction is necessary, weave one more row with the weft that is to be ended.

Close the slit by weaving from selvage to selvage, continuing with the established pattern. To judge how well the length and shape of the first and second halves of the cradle match, occasionally fold the sling at the midpoint. You may need to adjust your beat to correct the length. As you near the stage where the weaving tapers as the number of warp groups decreases from 8, to 4, to 2, you may need to tighten the weft in order to match the width of the first side.

Decrease the number of warp groups in stages in order to taper the end as follows:

**Step 1.** Regroup the threads so that there are 4 warp groups. Weave to match the other end.

**Step 2.** Regroup to two warp groups. Weave in a figure-of-eight to match the first end.

**Step 3.** Finally pull the warps together and bind all the threads tightly, tying off with a few half hitches. Do not cut the binding thread in case it is necessary to adjust and re-wrap the final binding after the second sling cord is complete.

FIGURE 7.27.
Partially woven cradle showing how diamond patterns and outlines are built.

## Making the Second Braid

The process to make this length of braid mirrors that of the first. It begins with the 24-strand diamond pattern with a core of 12, reduces to 24 strands, and finally to 16 strands. A split loop can be added near the end if desired. Reducing the number of strands means removing them from the braid. The point at which they are cut is indicated by an * in the diagrams below.

### Setting up the Stand for Making the Second Braid

**Stand.** Thread the first cord of the sling and the cradle down through the center hole of the braiding stand as seen in figure 7.28. To stabilize the warp before adding bobbins, insert a knitting needle on the underside of the stand just above the wrapping.

The twining will have maintained the original order of the first braid's ending. Slide the twining up so that it lies on the top of the stand to aid in attaching the bobbins. Add bobbins to all 32 strands. Untie the twining and the individual warp bindings. Place the eight red threads on the harness and arrange the brown and white threads on the braiding stand as shown in Step 1 of figure 7.29. Hang the counterbalance bag with its 49 oz. (1400 g) weight.

**Card.** Bring the warp through the hole in the card and arrange as seen in Step 1 of figure 7.29.

## Adding Warp D and Braiding the Diamond Pattern

Before braiding Steps 1–10, prepare the two red replacement strands of warp D. (Stand braiders add bobbins to the four ends.)

Add the first red warp on the diagonal, as shown in Step 1a of figure 7.29. (Card braiders place the warp in slots 9 and 25.) Follow Step 1b to lock it in place. Move the red threads to the harness/core. Work Step 2, add the second warp to the core in Step 2a. (Card braiders place the warp in slots 8 and 24.) Move the red threads (Step 2b). Work the diamond pattern by following Steps 3–20.

FIGURE 7.28.
Stand set up for braiding the second sling cord.

Work Steps 1 to 2b to add the red threads

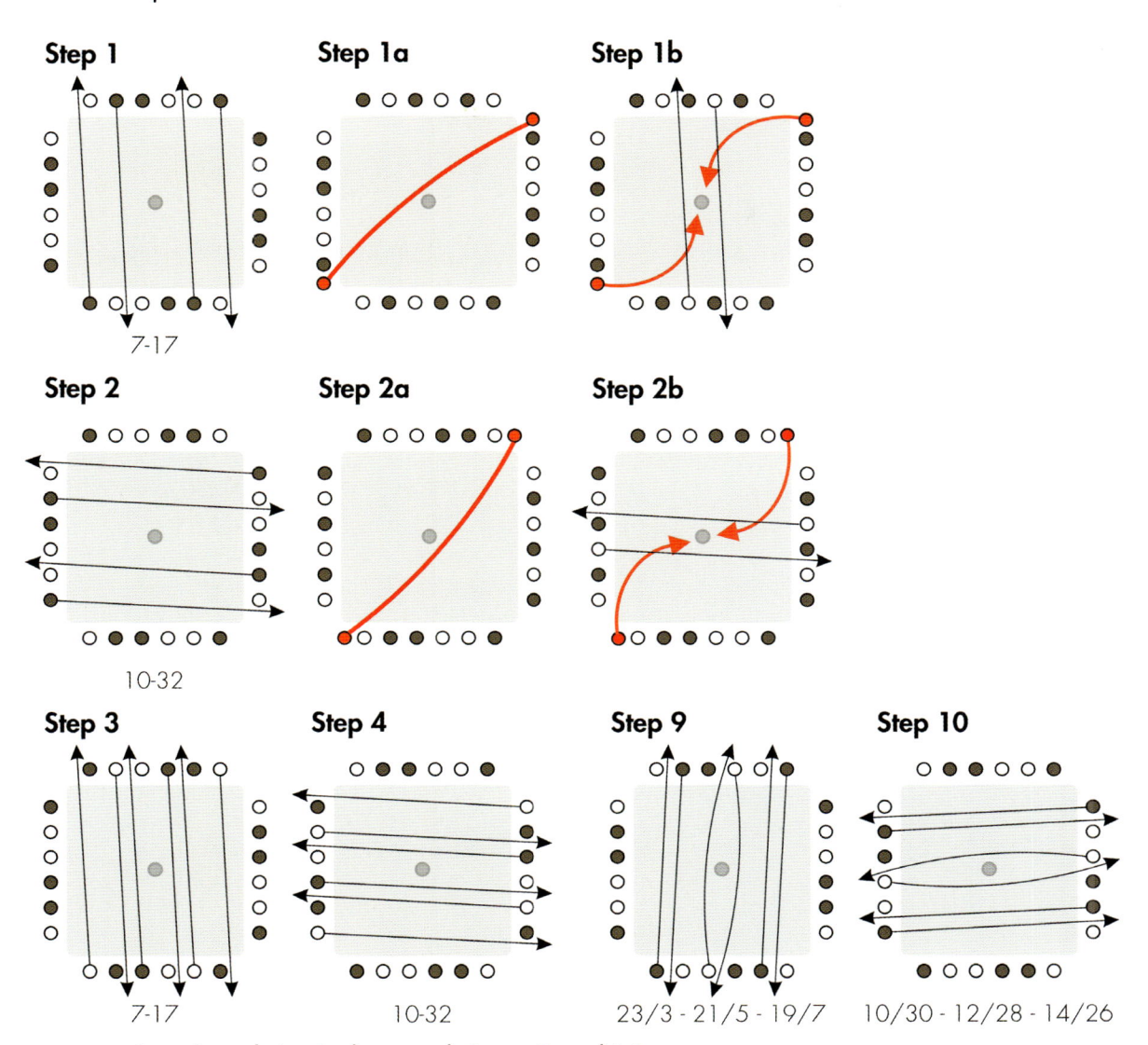

**Step 1**

7-17

**Step 1a**

**Step 1b**

**Step 2**

10-32

**Step 2a**

**Step 2b**

**Step 3**

7-17

**Step 4**

10-32

**Step 9**

23/3 - 21/5 - 19/7

**Step 10**

10/30 - 12/28 - 14/26

Repeat Step 3 and 4 x 3, then work Steps 9 and 10

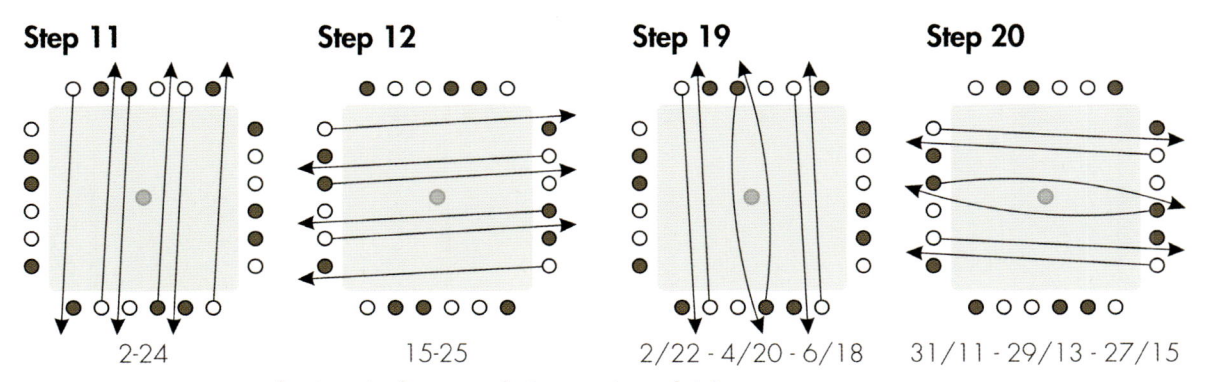

**Step 11**

2-24

**Step 12**

15-25

**Step 19**

2/22 - 4/20 - 6/18

**Step 20**

31/11 - 29/13 - 27/15

Repeat Steps 11 and 12 x 4, then work Steps 19 and 20

FIGURE 7.29.
Steps for adding the red strands that were cut before the cradle was woven and for beginning the diamond pattern.

## Completing the Diamond Pattern

Complete the brown and white diamond pattern by repeating Steps 21–40 twice. (See figure 7.30.)

Work Steps 61–70 in figure 7.31, replacing the white threads with red. An asterisk * indicates that white threads are to be cut close to point of braiding after finishing Steps 68, 72, and 80. (At the braider's discretion, the white threads can be removed more gradually by allowing them to be carried inside the core a little further. In this case, to create a sling with cords that are equal in diameter, the threads need to be added more gradually at the beginning as well.)

**Stand:** Note how the threads in Steps 63a and 64a are exchanged. Bringing the four white threads to the inner hooks aligns them in the center of the braid ready to be cut at the point of braiding after completing Step 68*. Threads at Steps 65 and 66 will be moved to the middle hooks and will be cut after completing Step 72*.

**Card:** When exchanging the white threads at Steps 63a and 64a, remember to "trap" the threads with the red threads being taken from the core. (See figures 5.19 and 5.20.) Pull them toward the center of the core. Then tie one knot in the end of each thread to mark it for cutting later. Mark the four white threads exchanged in Steps 65 and 66 with two knots each, and mark the last set of threads with three knots. After Step 68*, cut the four threads with single knots close to the point of braiding. Do the same for the core threads after Steps 72* and 80*.

## Reducing the Braid from 24 Strands to 16 Strands

At Steps 77 and 79, sets of four threads are lifted onto the harness, reducing the number of active threads to 16. After Step 80, the colors are in position to begin Design 16.3.1. (See figure 6.20.) The threads in Steps 77 and Step 79 will be gradually removed as the 16-strand braid is made. Cut four threads after eight steps and another four after four more steps.

**Stand**: Lift the threads indicated in Step 77 to the inside hooks of the harness. At Step 79 lift the threads to the middle position of the harness. Work eight steps of Design 16.3.1 and then cut the four threads from the center hooks. Braid four more steps and cut the remaining four threads. Reduce the counterbalance weight to 16 oz. (450 g).

**Card:** Move the four threads indicated in Step 77 to the core. Tie one knot in the end of each thread. Move the threads indicated in Step 79 to the core. Work eight steps of Design 16.3.1 and then cut the first set of four threads close to the point of braiding. Work four more threads and cut the remaining four threads.

Braid 16.3.1 to match the length of the first cord, ending with a tassel. (See *Finishing with a Tassel*, page 61, in Chapter 5.) If a split loop is desired, begin working it 6" (15 cm) before the end of the braid. (See Chapter 6, *Making a Split Loop*, page 139.)

**Repeat Steps 21 to 40 x 2**

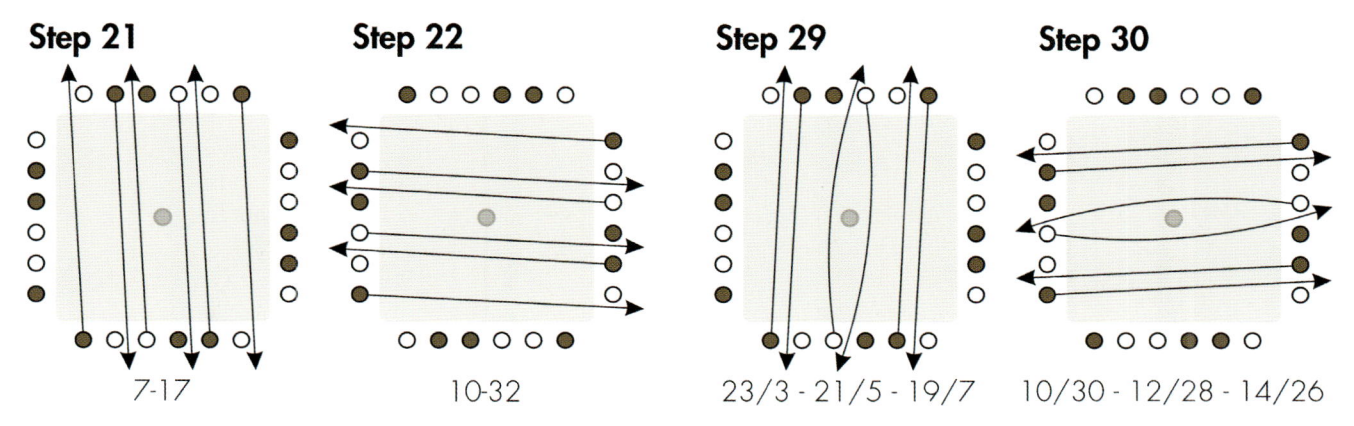

**Step 21**

7-17

**Step 22**

10-32

**Step 29**

23/3 - 21/5 - 19/7

**Step 30**

10/30 - 12/28 - 14/26

Repeat Step 21 and 22 x 4, and then work Steps 29 and 30

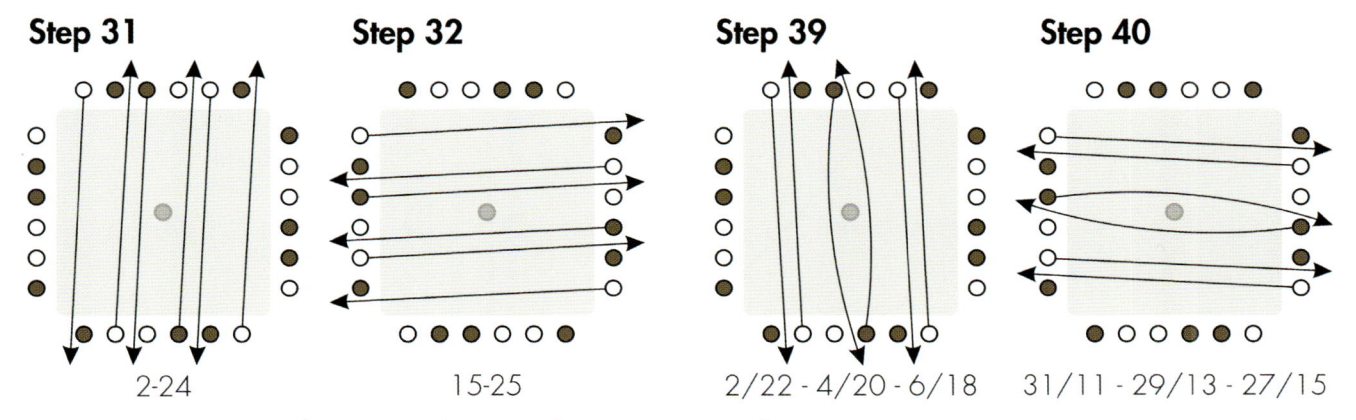

**Step 31**

2-24

**Step 32**

15-25

**Step 39**

2/22 - 4/20 - 6/18

**Step 40**

31/11 - 29/13 - 27/15

Repeat Step 31 and 32 x 4, then work Steps 39 and 40

FIGURE 7.30.
Complete the brown and white diamonds by repeating Steps 21–40 twice, bringing the total number of Steps completed to 60.

**Step 61**

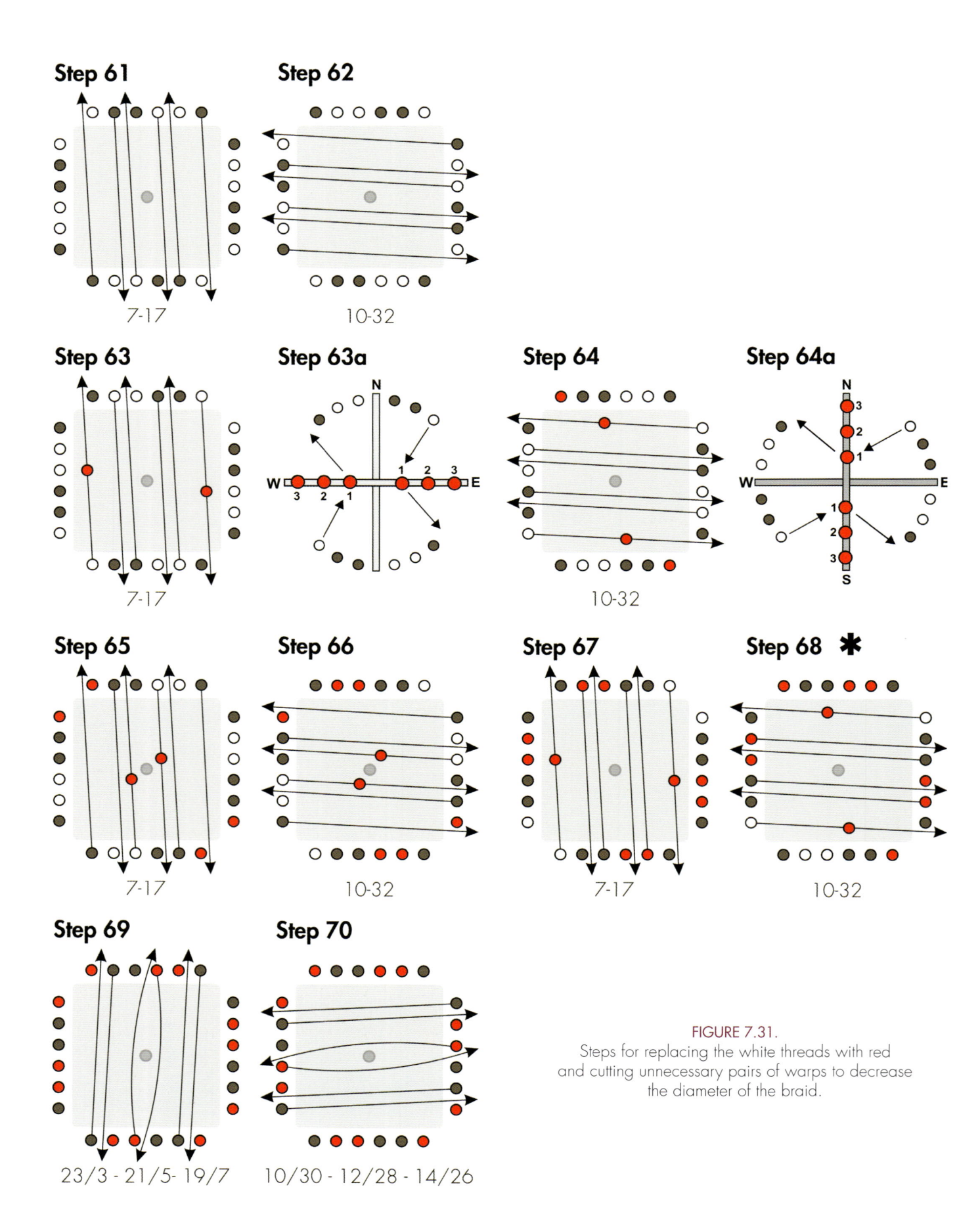

7-17

**Step 62**

10-32

**Step 63**

7-17

**Step 63a**

**Step 64**

10-32

**Step 64a**

**Step 65**

7-17

**Step 66**

10-32

**Step 67**

7-17

**Step 68** ✱

10-32

**Step 69**

23/3 - 21/5 - 19/7

**Step 70**

10/30 - 12/28 - 14/26

FIGURE 7.31.
Steps for replacing the white threads with red and cutting unnecessary pairs of warps to decrease the diameter of the braid.

**Step 71**

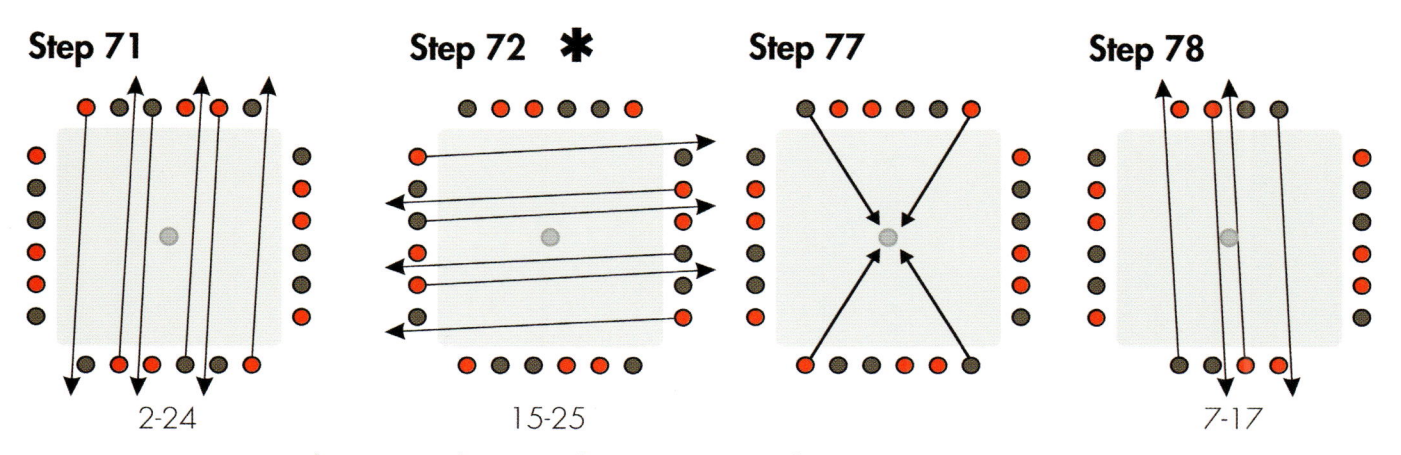

2-24

**Step 72** ✱

15-25

**Step 77**

**Step 78**

7-17

Repeat Step 71 and 72 x 3, then work Steps 77 and 78

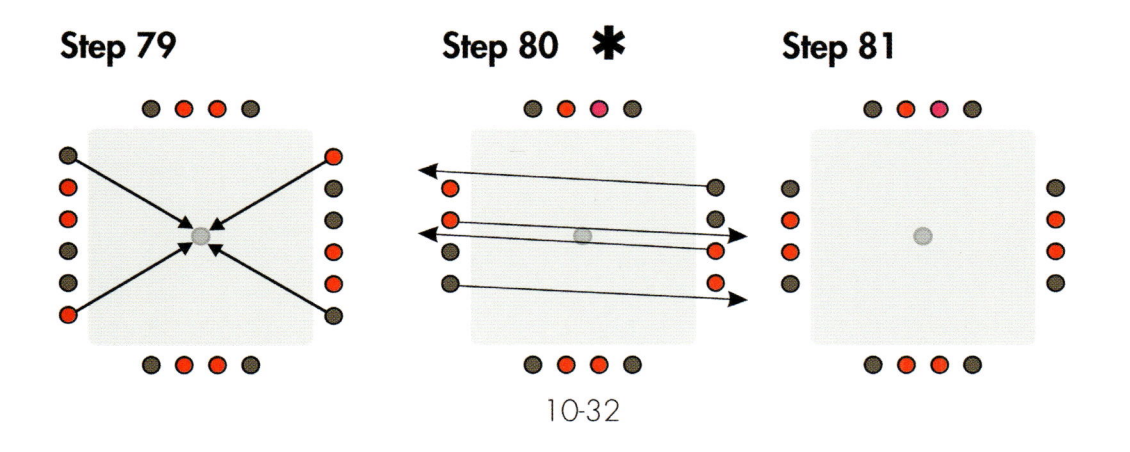

**Step 79**

**Step 80** ✱

10-32

**Step 81**

FIGURE 7.32.
Steps for reducing the 24-strand braid to 16 strands.

## Additional Sling Construction Methods

The following stories illustrate two other methods of constructing slings. Laverne Waddington, years before becoming an expert in backstrap weaving, had the opportunity to make a sling under the tutelage of a local sling maker while visiting Peru. Laverne gives a clear description of a traditional method for making a braid in the hand. Her sling was constructed in the Macusani style. (See figure 7.33.) Ben Turner makes slings that are both attractive and effective for slinging. In his search to improve the accuracy and durability of his slings, he takes a non-traditional approach, drawing from Tibetan, Māori, and Andean textile techniques. (See figures 7.34 and 7.35.)

### Laverne Waddington's Story

While on my second trip to Peru in 1997 looking for weaving teachers, I met Zacarias in Yanque, in the Colca Canyon [Arequipa Region]. He was farming the land and taking care of the llamas that were part of a small hotel at the bottom of the canyon.

I had bought some slings in a store in neighboring Chivay and Zacarias named the braiding styles for me . . . *Salsa, Cajamarca, Palma,* and *Margarita,* pointing out the Cajamarca as the most difficult technique and the only one that he was unable to do. [Margarita is a diamond pattern similar to Design 24.10.] I really wanted to learn the Margarita braid but Zacarias insisted that we should start out with the easier Palma. It was the time of the barley harvest and so Zacarias would show me the moves to get me started in the morning and then leave me to go work in the fields.

We used black, brown, and white llama fiber which had been hand spun and plied by his wife. The first braided side of the sling was made almost entirely in the 24-strand Palma technique. Four strands of each color were measured from the nose to the toes and then doubled around a thick cord, which was held in the fist.

Zacarias showed me how to tilt my wrist and let the strands, which ended in butterflies, hang in such a way that they would be less likely to spiral around each other and get entangled. This position also made the uppers and lowers much more clearly visible. The vocabulary was simple, which was just as well, as my Spanish was not up to much in those days . . . *cerca, lejos, arriba, abajo* which referred to the strands on the near side and far side of my wrist and the uppers

**FIGURE 7.33.**
Sling with Palma braid cords and tapestry woven cradle by Laverne Waddington. *Courtesy of Laverne Waddington.*

and lowers. In any case it was more about watching than listening to instructions.

After braiding for about 27 inches, the 24 strands were split into two groups to form the finger loop. The two sides of the finger loop were made from a nameless 8-strand 2-color braid with the 4 strands of the third color acting as the core. The strands were then rejoined to finish the last inch or so in the Palma technique.

The cradle section of the sling was woven in weft-faced tapestry technique. The thick cord was withdrawn from my Palma braid and replaced with 12 new doubled strands which would form the warp of the woven cradle and the braid on the other side. The strands were coiled around a stick to spread them to the desired width and tensioned between my body and a tree. Zacarias wove the first motif, split the warps into two halves to form the slit and then left me to continue and duplicate his motif on the other end.

He taught me to make the braid on the other side of the cradle in the Palma technique for about 8 inches and then finish with the finger-loop braid, working two colors for 2 inches before replacing one of the colors with the core color and continuing. He made a point of telling me that fringes and tassels on his slings were to be kept to a minimum and he was not at all keen on some of the quite flamboyant decorations that new braiders were using these days! He showed me how to sew a two-color edging around the cradle using black and white acrylic yarn. I think he was pleased with my work as he then agreed to teach me the Margarita braid.

## Ben Turner's Story

What influenced me to start making slings? A desire to sling! When I became interested in slinging, there was very little information available (even on the Internet) about how slings were made and buying them wasn't really an option, so if I wanted to sling I had to make my own. Making a sling may seem straight forward, such as tying some boot laces to a piece of leather, but actually things like the correct length of the sling, the size and shape of the pouch, the type of projectile you intend throwing (i.e., a lead fishing sinker, a golf ball, or a tennis ball), and choice of materials along with your preferred slinging style, all come into play.

If you have never slung before and your main point of reference is an artist's interpretation of David v. Goliath, it's a lot of trial and error to make a sling that is functional. I had long had an interest in knot tying and was already a member of the IGKT [International Guild of Knot Tyers] so it was a natural progression from making knotting projects to sling making. After viewing pictures and resources shared on http://slinging.org of Tibetan- and Andean-style slings, using some educated guesswork, I was able (poorly at first) to make those types of slings with the techniques and processes used in the different regions.

I prefer to make twined pouches because of a long-standing interest in Maori Taniko weaving. Twined pouches are simple to make and allow for a variety of patterns to be made easily. Twined pouches also allow a slinger to use any size or style of projectile.

FIGURE 7.34.
Sling made by Ben Turner featuring reversing and striped patterns in the sling cords and a twined cradle. *Courtesy of Ben Turner.*

FIGURE 7.35.
The braids in Ben's second sling feature a variety of Andean/Tibetan 16-strand braids with a Tibetan-style twined cradle. *Courtesy of Ben Turner.*

The sling in figure 7.34 was made with a 14-ply wool, acrylic and alpaca yarn, approximately 55–63 yds. (50–58 m) per 1.75 oz. (50 g) ball. I find the 14-ply thickness just perfect for Andean/Tibetan braiding. The zigzag pattern was made with alternate rotations at each step, it was also used for the straight braid, and both braids were made with 16 strands.

The pouch was twined with the basic twining technique, although, in this case, the weft consisted of three separate 8-ply yarn strands used as one weft (by folding them in half), making 2 wefts of 3 ends each twining around each other, this gives the pouch strength and body. The strands near the end of the pouch are whipped for strength and chafing protection from abrasive rocks, etc.

I made the pattern in the pouch and arranged the colors this way for the aesthetic appearance. The finger loop was designed to fit over two fingers as this is the grip I find most comfortable for me and offers the greatest amount of control. The round bit on the finger loop is half hitching and the tassel was added for decorative effect.

In the sling with the black and white pouch [see figure 7.35], the Andean/Tibetan braids used were (in Cahlander's nomenclature) II.B.6 and II.A.1, both 16 strands. On the other side of the sling is braiding (16 strands) around a core that comes into three 8-strand braids that were stitched together, for a total of 24 strands. The pouch is twined with four strands, two dark and two light, in two different twining techniques in order to produce the pattern that breaks up the sling pouch into quadrants. Lastly, the tassels and wrappings are formed by combing out 3-strand wool and securing with basic whippings (simple/west country, etc.). The wrappings are also half-hitched.

FIGURE A.1.
Core frame designed by Doug Newhouse.

# APPENDIX A
## Plans for a Core Frame

The braiding stand sits inside the core frame beneath the harness as shown in figure A.1. The harness is made with two crosspieces that form the four arms. Each arm has three hooks and a slot for holding the core threads. The crosspieces are attached to a dowel that can be moved vertically to adjust the working height of the core bobbins. The height of the braiding stand should be adjusted to work with the height of the braider's chair. As a general rule, the braider's forearms are roughly parallel to the floor when braiding.

The core frame, pictured in figure A.1, allows a low braiding stand to be raised for working from a chair. For those building for a tall stand, the legs on the core frame base can be shortened and the upright that supports the cross arm and harness can be lengthened.

## Dimensions

The essential measurements are shown in figure A.2. While it is not possible to give universal measurements that would fit all braiding stands, there are two measurements that are essential:

1. The distance between the top edge of the braiding stand and the core frame's upright must be at least 5" (125 mm). This gives sufficient clearance so that the bobbins do not strike the upright as they are moved.

2. The distance between the top of the braiding stand and the underside of the cross arm should not be more than 12" (305 mm). (See figure A.2.)

3. The dimensions of the recess that is inside the core frame base are determined by the measurements of the braiding stand's base.

4. The distance from the front edge of crossarm (B) to the center of the harness support hole is 1" (25 mm).

5. The holes in the harness support dowel are ½" from center to center. (In most cases only the first few holes will be used.)

## Core Frame Joinery

The join that secures the crossarm to the upright must be strong. The suspended harness must be able to carry up to 6.6 lbs. (3 kg) when using 70 g bobbins to make a 24-strand braid with a core of 16 strands. The legs and base are secured with mortise and tenon joints. (If you wish to make a stand that packs easily for traveling to workshops, the upright can be made with two pieces that are joined with a half-lap splice joint fastened with a nut and bolt.)

The harness consists of two crosspieces that are lap jointed and attached to the dowel with a wood screw. A series of evenly spaced holes are drilled in the dowel to make it adjustable. A small dowel is inserted to hold the harness assembly in place. (See figure A.3.)

On the underside of each arm, three hooks are equally spaced. The spacing of 1" (25 mm) between the hooks on our model is based on the dimensions of the 70 g bobbins made by BraidersHand. They are 1½" (38 mm) high by 1¼" (32 mm) wide. *Since the spacing controls the angle of the core threads, avoid lengthening the arms of the harness.* Lengthening would increase this angle causing the structure to distort. No adjustment is needed if smaller bobbins are used with this spacing. A groove is cut on the top of each arm, *close to the dowel*, to hold the extra four threads in braids with 16-strand cores.

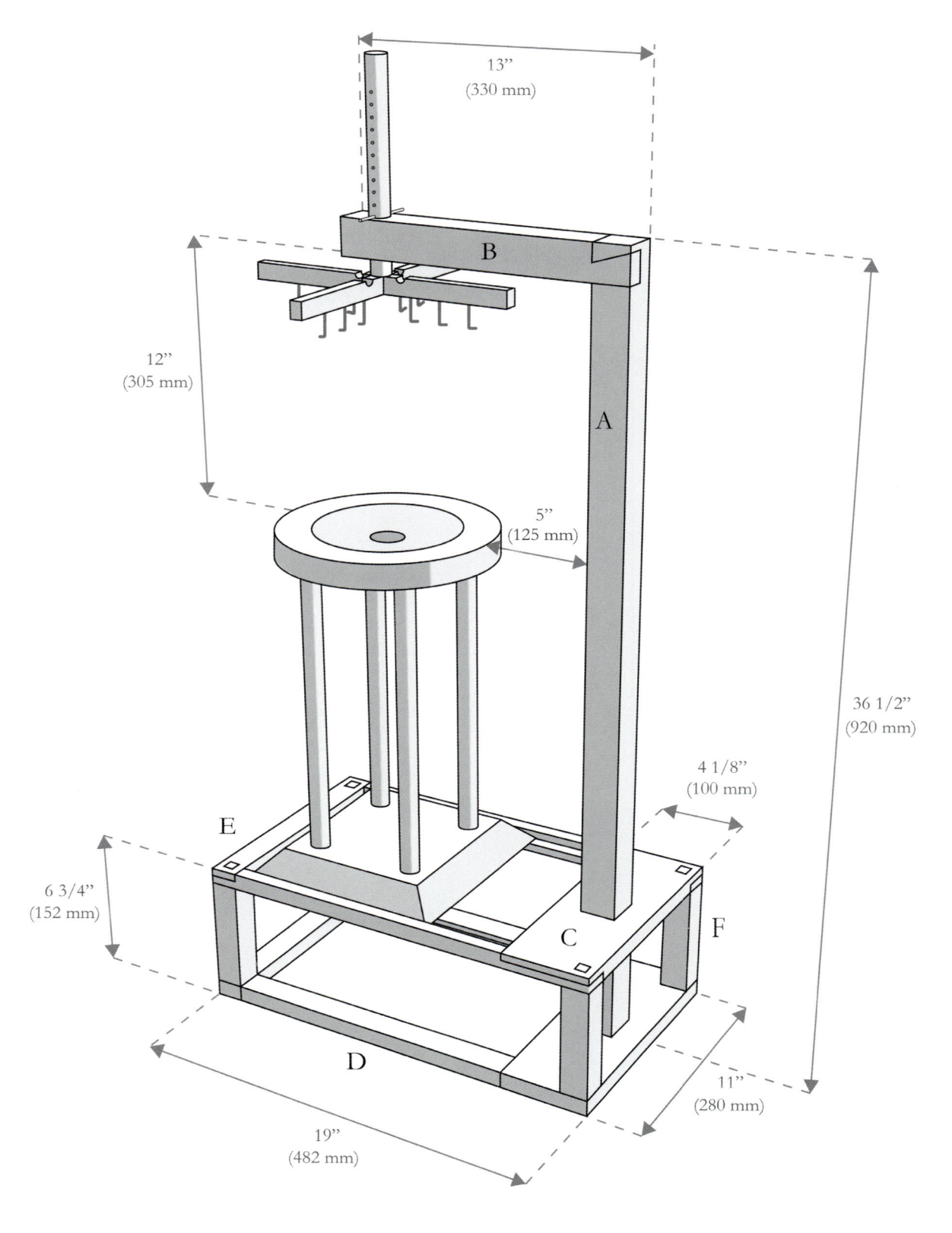

FIGURE A.2.
Core frame plans.

12"
(304 mm)

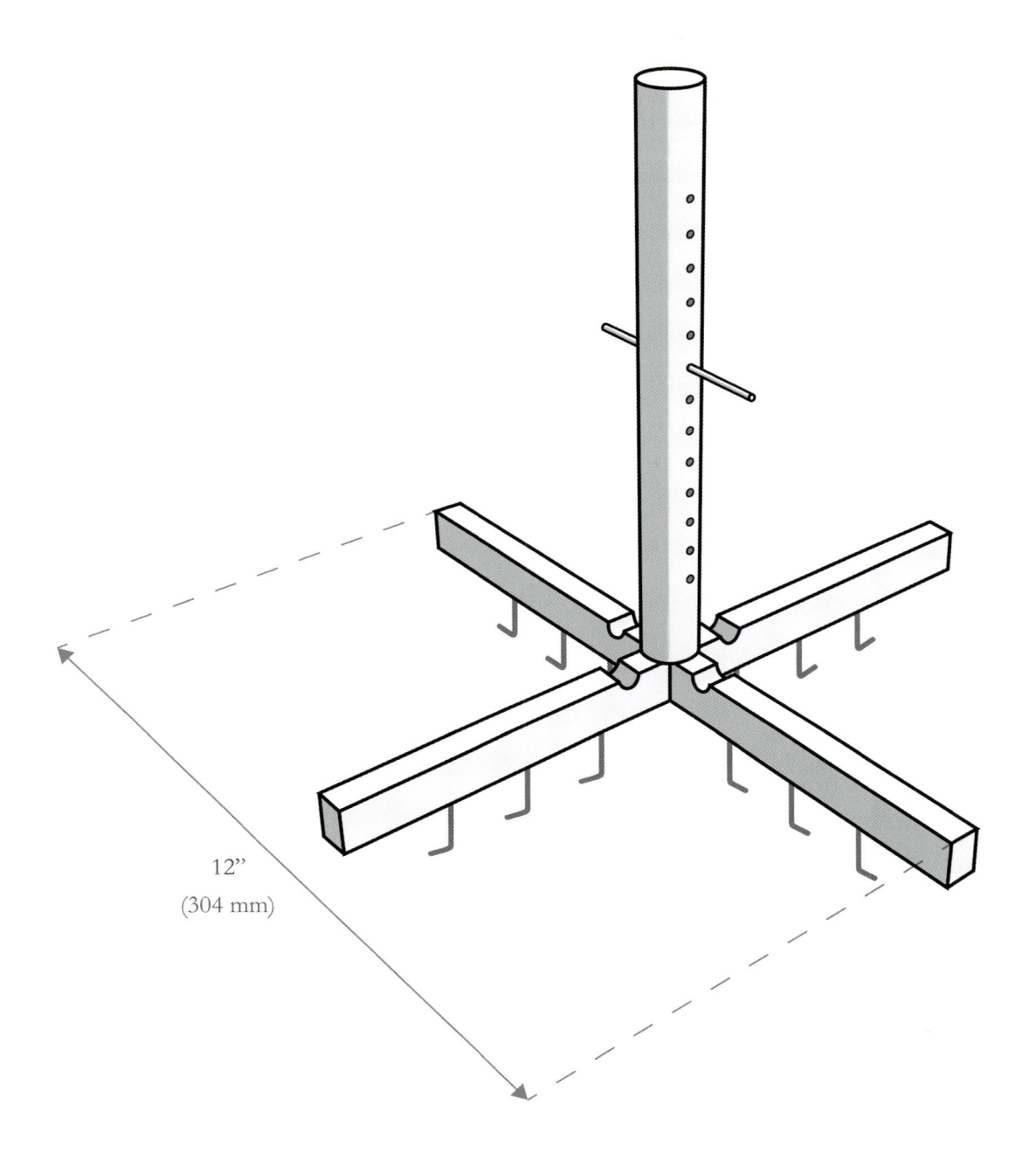

**FIGURE A.3.**
Harness plans.

## Materials Needed

This core frame has been designed to be easy to transport and assemble. It fits together with only two metal fasteners, a wood screw that attaches the harness to the dowel, and a bolt and wing nut that fastens the pieces of the upright together.

Because of the weight that it will carry, the core frame should be made from a close-grained hardwood such as maple, beech, sycamore, etc. Figure A.4 shows the finished and rough sizes of the frame components. If you decide to make the upright in two pieces so that it disassembles for travel, you will also need a nut and bolt to fasten it together.

## Harness

For the harness you will need:

- Harness arms: two 12" × 3/8" × 1/2" (19 mm × 10 mm) pieces of scrap hardwood

- 9" (228 mm) length of 7/8" (22 mm) hardwood dowel for harness support

- 12 "L"-shaped brass screw hooks that extend 1" (25 mm) from the screw. (Bend slightly to make the angle a little more acute. See figure A.3.)

- 4" (10 mm) length of 3/16" (5 mm) diameter dowel to secure the harness support to the crosspiece

- 1 woodscrew for securing dowel to crosspieces

- 3/16" (5 mm) drill bit for drilling holes in dowel

- 7/8" (22 mm) drill bit for drilling the hole in the crossarm

| | Components | Finished size | Metric | Rough size |
|---|---|---|---|---|
| A | upright | 1 off 36½" × 1¾" square | (93 cm × 45 mm) | 1½ board feet 8/4 hardwood |
| B | crossarm | 1 off 13" × 1¾" square | (33 cm × 45 mm) | |
| C | anchor board | 2 off 11" × 4" × ¾" | (28 cm × 10 cm × 19 mm) | 1 board foot 4/4 hardwood |
| D | rails | 4 off 19" × 1¼" × ¾" | (48 cm × 32 mm × 19 mm) | |
| E | frame ends | 2 off 11" × 1¼" × ¾" | (28 cm × 32mm × 19 mm) | |
| F | posts | 4 off 6¾" × 1 ¼" square | (17 cm × 32 mm) | 1 board foot 6/4 hardwood |

FIGURE A.4.
Finished sizes of the components that make the core frame.

# APPENDIX B
## Plans for Making a 32-slot Card for Core Braiding

It is recommended that the card be made from mat or illustration board. The durability of the card will be improved by gluing a paper template to the card. Print the template shown below in figure A.5 on card stock with a weight of 60–65 lbs. Use 160–170 gsm card in the UK. The foam plate used for Japanese kumihimo can be adapted for core braiding by increasing the size of the center hole and cutting extra slots on all four sides.

### Numbering and Cutting the Card

Cut a 4¼" (10 cm) square card. Print out the template found in figure A.5. Adhere the template to the card with a permanent adhesive (glue stick or spray adhesive). Smooth out any wrinkles.

If a template is not being used, measure and mark the position of the slots and the center hole according to the measurements shown in figure A.5. Use a craft knife to cut the square center hole so that it has a clean edge. Note that the center hole is larger than the hole normally needed for Japanese kumihimo.

Use scissors (not a knife) to cut the slots around the outside edge of the card. The cuts should be about ⅜" (1 cm) deep. Cut the slots using the throat of the scissors, not the points. Note the way the card below is numbered and the direction that the numbers face. If you are writing the numbers by hand, it is important to place them slightly to the left of each slot so they remain visible when working.

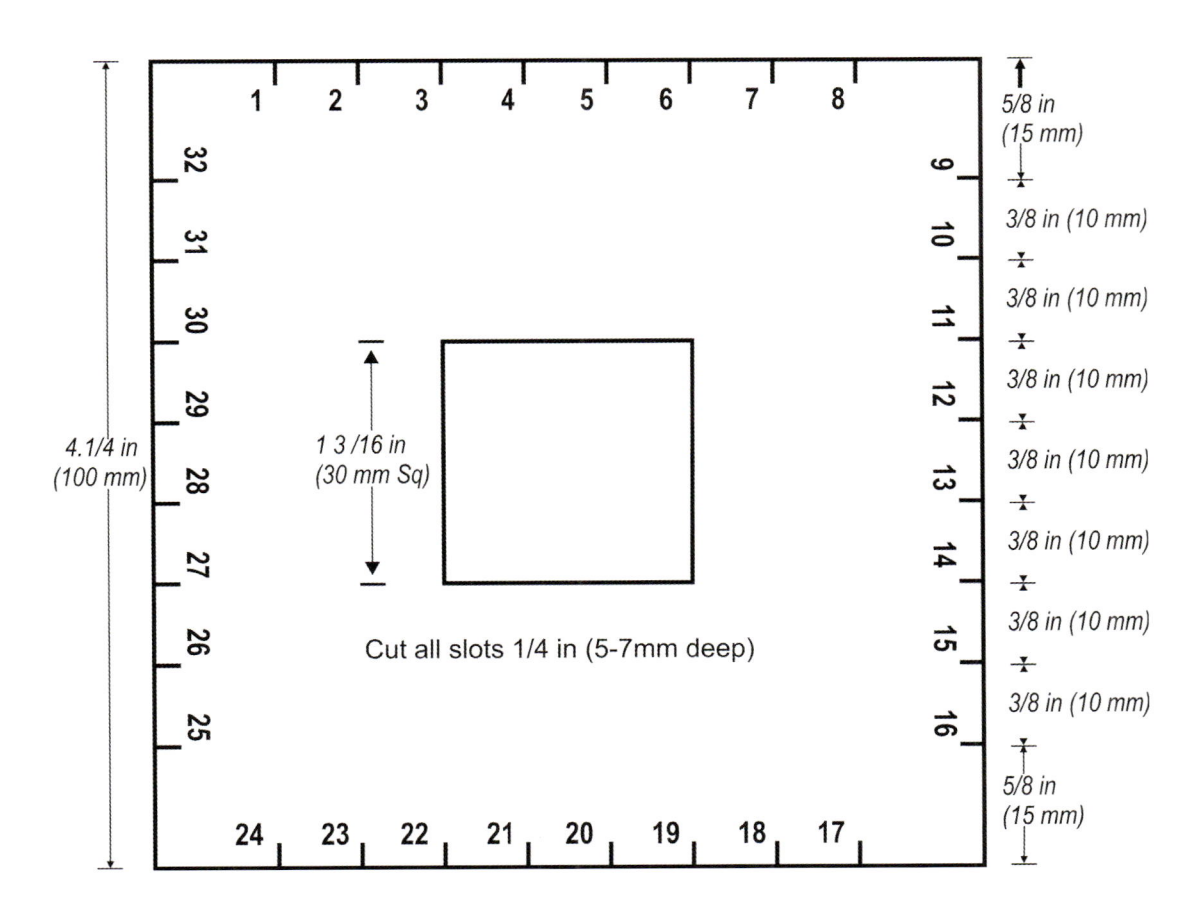

**FIGURE A.5.**
Measurements for making a 32-slot card for core braiding.

## Key Moves for Working on the Card

**24-Strand Rotational Moves**

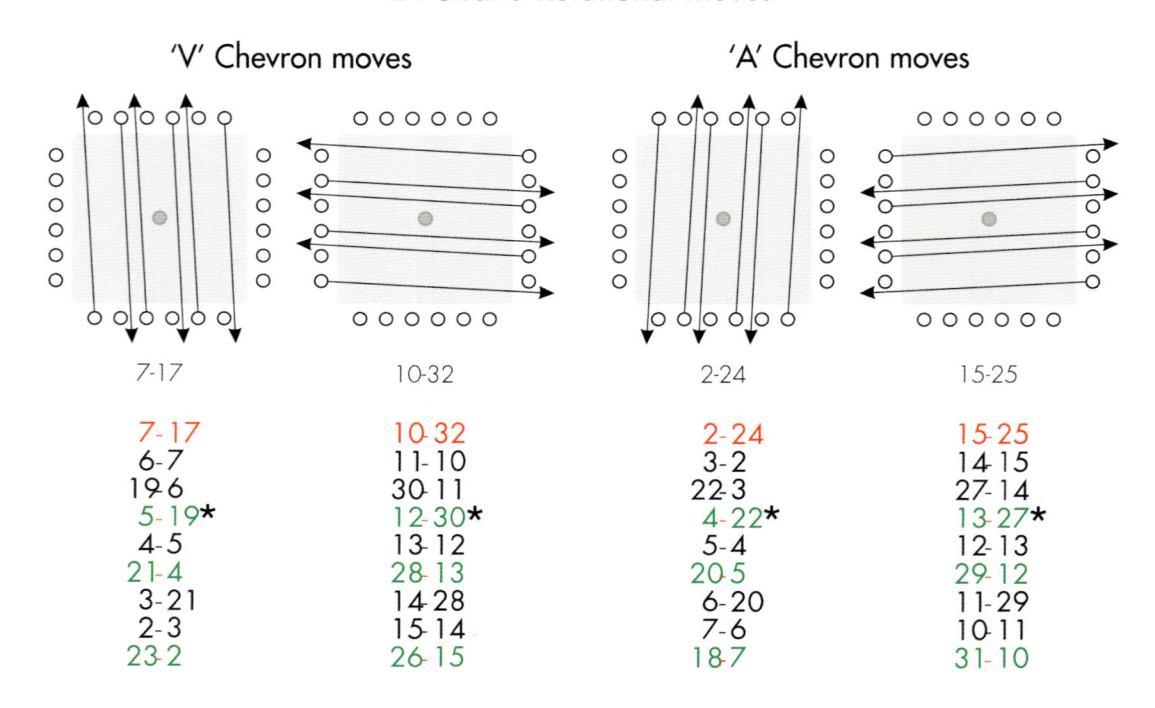

'V' Chevron moves            'A' Chevron moves

| 7-17 | 10-32 | 2-24 | 15-25 |
|---|---|---|---|
| 7-17 | 10-32 | 2-24 | 15-25 |
| 6-7 | 11-10 | 3-2 | 14-15 |
| 19-6 | 30-11 | 22-3 | 27-14 |
| 5-19* | 12-30* | 4-22* | 13-27* |
| 4-5 | 13-12 | 5-4 | 12-13 |
| 21-4 | 28-13 | 20-5 | 29-12 |
| 3-21 | 14-28 | 6-20 | 11-29 |
| 2-3 | 15-14 | 7-6 | 10-11 |
| 23-2 | 26-15 | 18-7 | 31-10 |

\* Flip core after this move

Note: On completing each column, reposition the threads to their set-up slots

**FIGURE A.6.**
Rotational moves for working on the 32-slot card.

## Plans for Making a 40-slot Card for Design 32.1

The card stock and cutting instructions for the 40-slot card shown in figure A.7 are the same as those given for the 32-slot card shown in figure A.5.

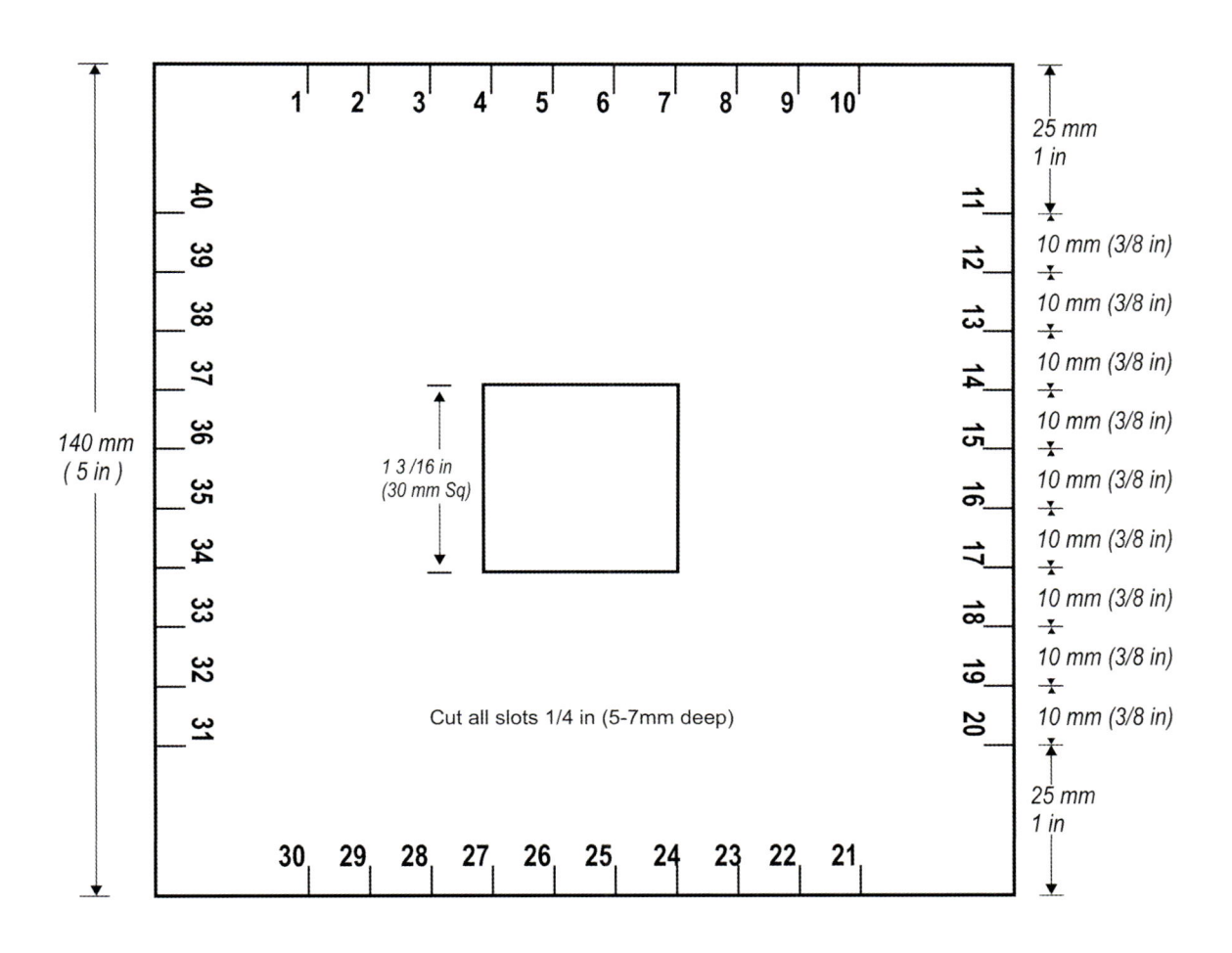

*Diagram shows a square card with dimensions 140 mm (5 in) tall. Slots numbered 1–10 along the top, 11–20 down the right side, 21–30 along the bottom, and 31–40 up the left side. A 30 mm square (1 3/16 in Sq) is cut out of the center. Right side margins: 25 mm (1 in) from top to slot 11, then 10 mm (3/8 in) between each slot 11 through 20, then 25 mm (1 in) from slot 20 to bottom. Instruction: "Cut all slots 1/4 in (5–7mm deep)".*

**FIGURE A.7.**
Measurements for making a 40-slot card for Design 32.1.

# APPENDIX C
## Rotational Move Diagrams for 24-strand Core Braids Methods 1 and 2

Two methods of working are shown in the four diagrams A.8–A.11.

*Method 1*, done with the braider standing, starts with moving pairs from the same side of the stand, e.g., two S threads move N, two N threads move S. The last move is made with a center N/S pair. See figures A.8 and A.9.

*Method 2* can be worked sitting or standing. The threads are moved from opposite sides of the stand, diagonally across from each other.

Try both methods. As you begin to understand how the threads move, you may find another variation that also works for you.

### Method 1 - V shaped rotational moves (N/S clockwise, E/W anticlockwise)

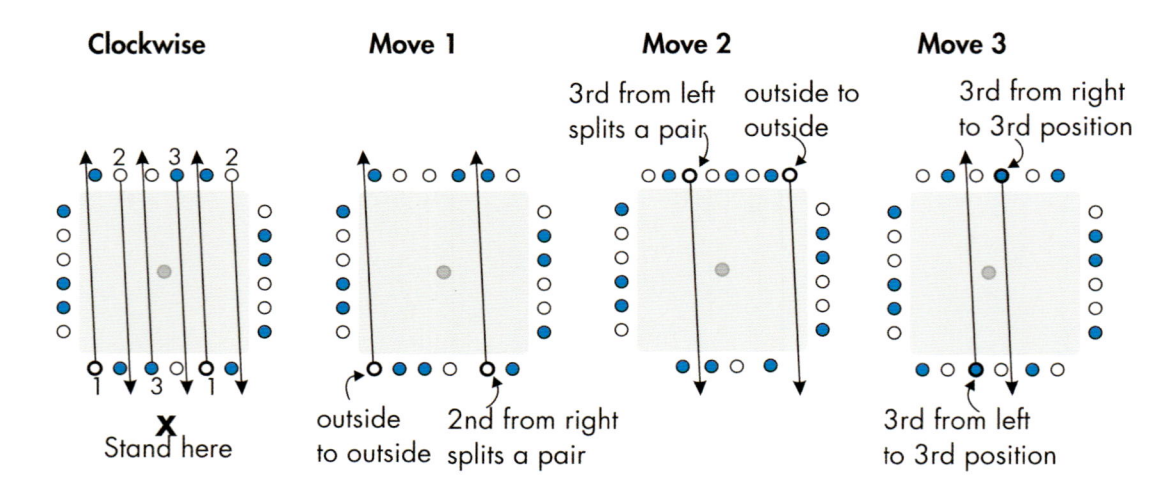

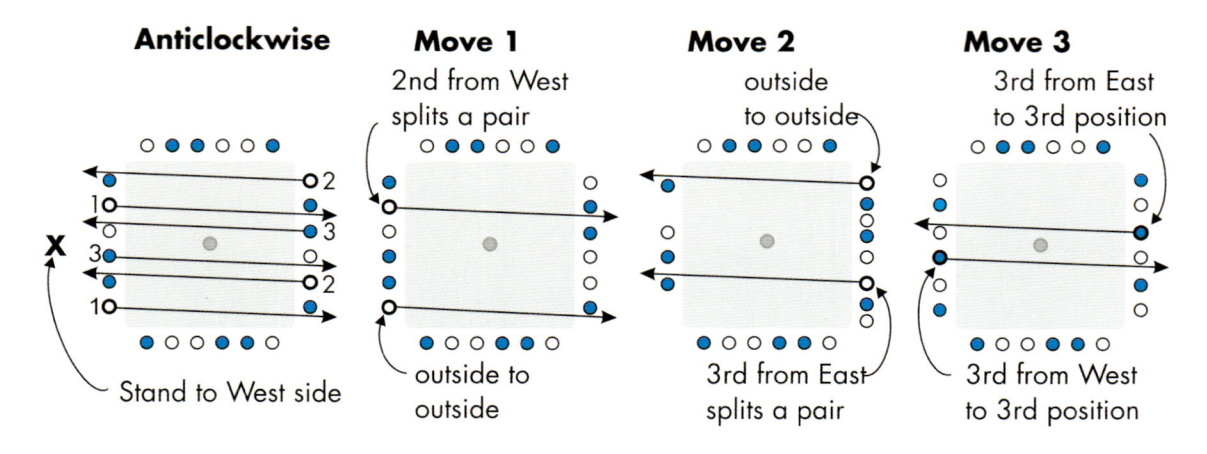

**Clockwise** — Stand here

**Move 1** — outside to outside; 2nd from right splits a pair

**Move 2** — 3rd from left splits a pair; outside to outside

**Move 3** — 3rd from right to 3rd position; 3rd from left to 3rd position

**Anticlockwise** — Stand to West side

**Move 1** — 2nd from West splits a pair; outside to outside

**Move 2** — outside to outside; 3rd from East splits a pair

**Move 3** — 3rd from East to 3rd position; 3rd from West to 3rd position

FIGURE A.8.
Method 1 for braiding a *V-shaped* pattern begins with braiding pairs from the same side of the stand (Japanese-style). It is worked standing.

**FIGURE A.9.**
Method 1 for braiding an *A-shaped* pattern begins with braiding pairs from the same side of the stand (Japanese-style). It is worked standing.

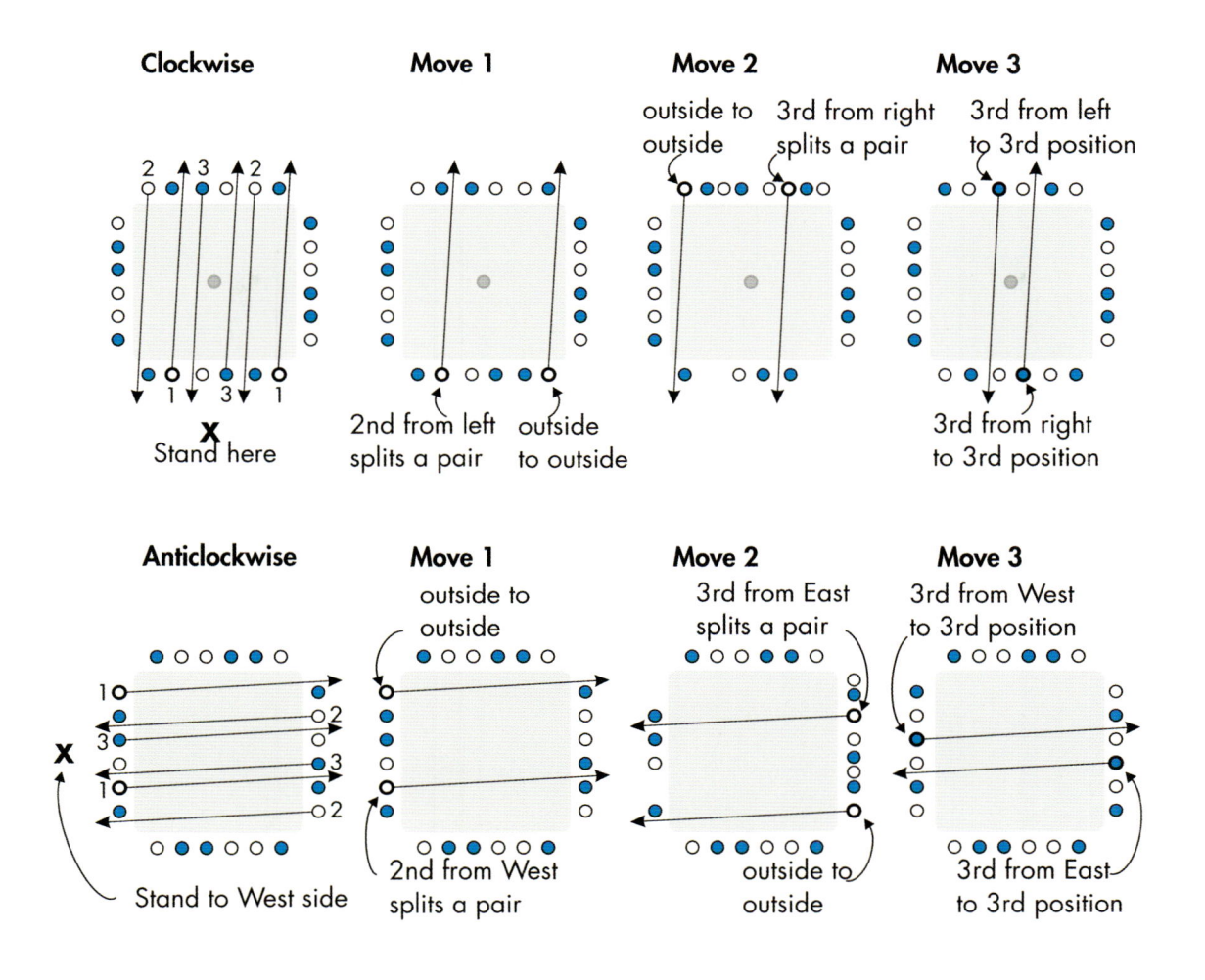

## Method 2 - V shaped rotational moves (N/S clockwise, E/W anticlockwise)

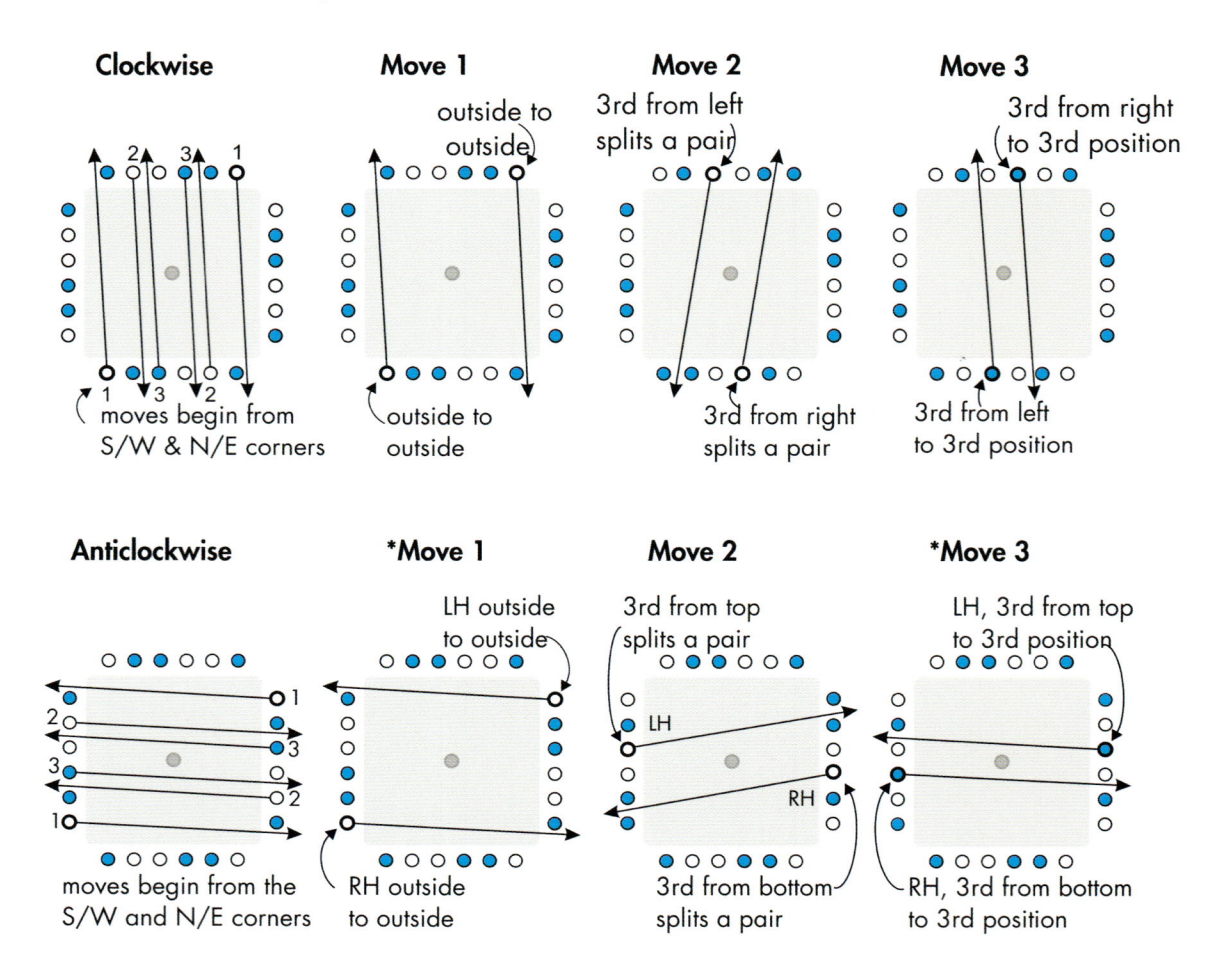

**Clockwise** — moves begin from S/W & N/E corners

**Move 1** — outside to outside

**Move 2** — 3rd from left splits a pair; 3rd from right splits a pair

**Move 3** — 3rd from right to 3rd position; 3rd from left to 3rd position

**Anticlockwise** — moves begin from the S/W and N/E corners

**\*Move 1** — LH outside to outside; RH outside to outside

**Move 2** — 3rd from top splits a pair; 3rd from bottom splits a pair

**\*Move 3** — LH, 3rd from top to 3rd position; RH, 3rd from bottom to 3rd position

\* In Moves 1 & 3 the left hand reaches behind the core to pick up the west thread, the right hand reaches in front to pick up the east thread, and then they are placed. In Move 2 the hand position is natural.

FIGURE A.10.
Method 2 for making a *V-shaped* pattern is worked with diagonal pairs.
It can be made sitting or standing.

## Method 2 - A shaped rotational moves (N/S anticlockwise, E/W clockwise)

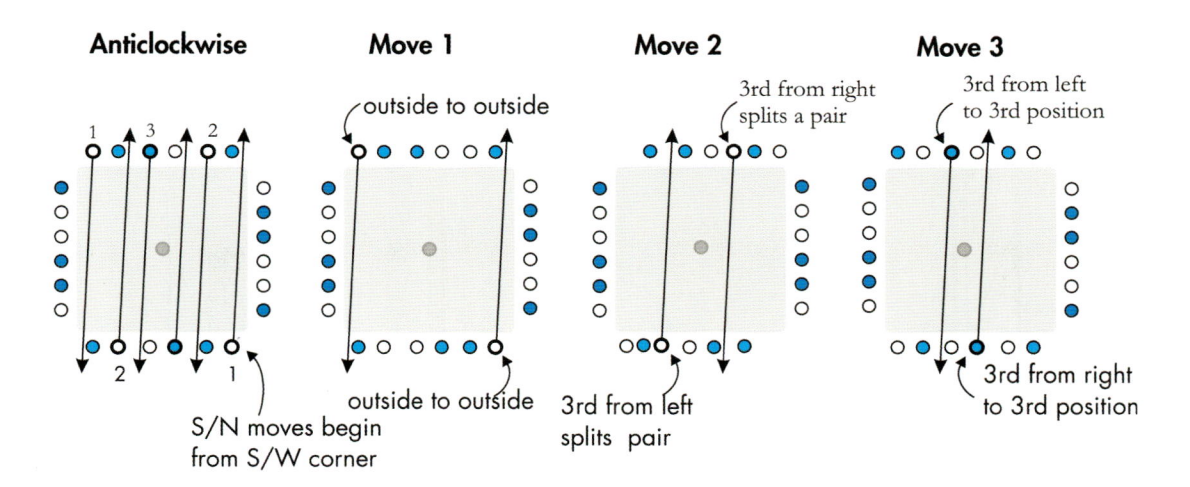

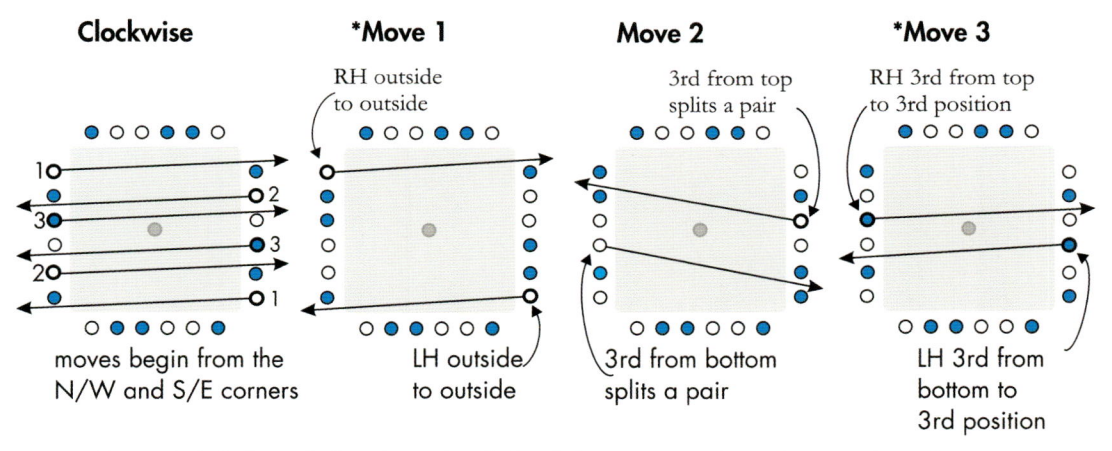

**Anticlockwise**

S/N moves begin from S/W corner

**Move 1**: outside to outside — outside to outside

**Move 2**: 3rd from right splits a pair — 3rd from left splits pair

**Move 3**: 3rd from left to 3rd position — 3rd from right to 3rd position

**Clockwise**

moves begin from the N/W and S/E corners

***Move 1**: RH outside to outside — LH outside to outside

**Move 2**: 3rd from top splits a pair — 3rd from bottom splits a pair

***Move 3**: RH 3rd from top to 3rd position — LH 3rd from bottom to 3rd position

\* In Moves 1 & 3 the right hand reaches behind the core to pick up the west thread, the left hand reaches in front to pick up the East thread, and then they are placed. In Move 2 the hand position is natural.

FIGURE A.11.
Method 2 for making an *A-shaped pattern* is worked with diagonal pairs.
It can be worked sitting or standing.

# APPENDIX D
## Design Planning Diagram

We have included a blank planning diagram for braiders that can be used in a variety of ways. Braiders may find that it is sometimes helpful to redraw designs when they are using colorways that are quite different from the designs shown in the book. Some braiders will find it helpful to draw each move of a step when they are initially learning to make core braids. And some will want to record new designs and variations.

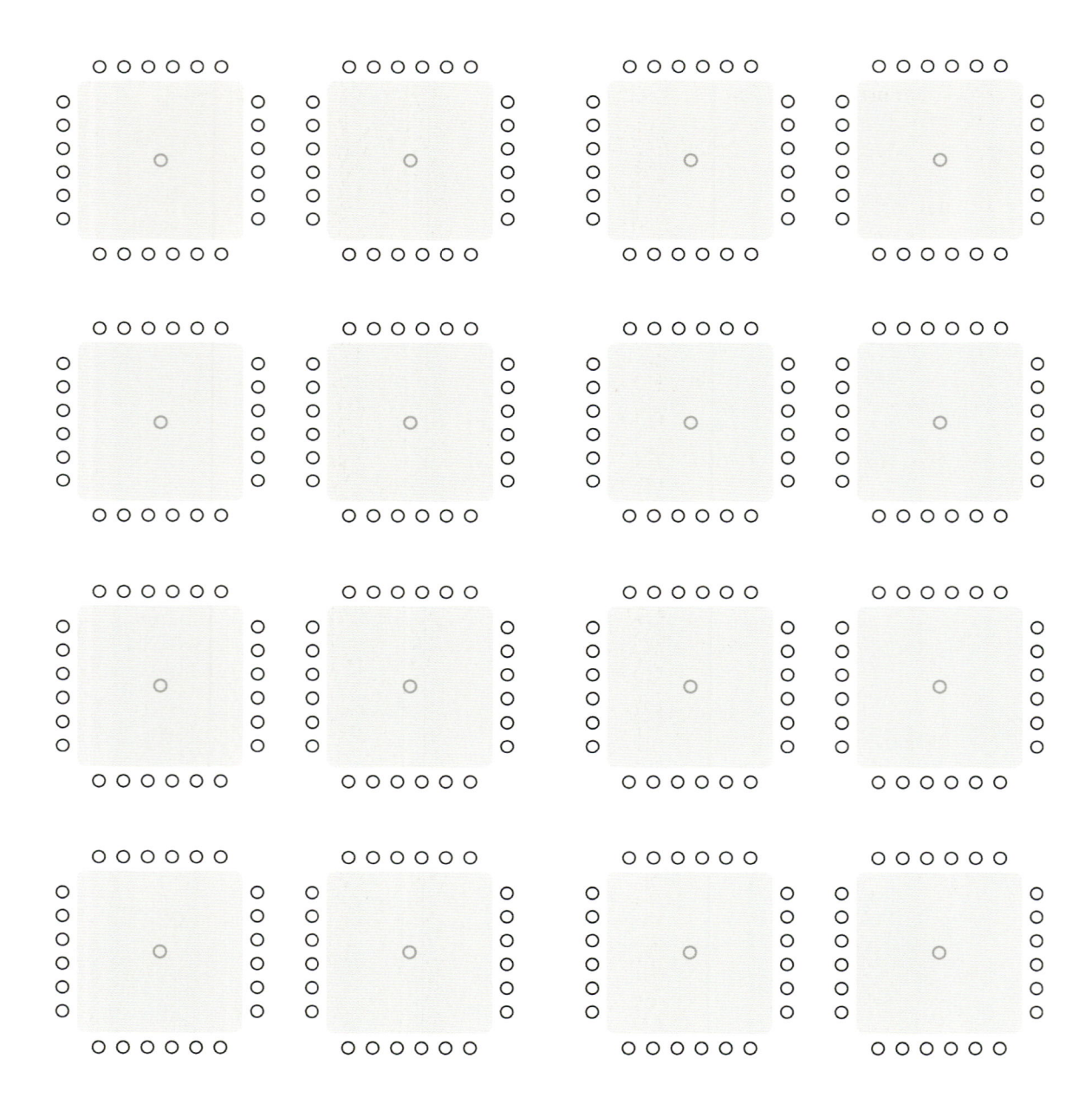

FIGURE A.12.
Planning diagrams for recording information about new patterns or design variations.

# APPENDIX E
## Reading Core Exchange Diagrams

Figure A.13 shows the familiar Step diagram with the addition of a red arrow pointing to a yellow circle in the center of the thread-movement arrow. The circle indicates that an orange thread is to be exchanged for a yellow thread from the core.

Figure A.14 is a bird's eye view of the yellow core threads arranged on the E/W arms of the harness with the orange and brown threads below on the N/S sides of the stand. The core threads on the harness are marked 1, 3, and 2 indicating the order in which the threads will be exchanged. (Although moving the core threads in this order seems illogical on paper, in practice it provides a smoother transition for threads moving between the braid and the core.) Figure A.15 shows the placement of the stitches on the braid.

### Steps for Exchanging the Core on the Stand

To change Design 24.2.1a's brown and orange chevron pattern to brown and yellow, follow Steps 1–6 detailed in figures A.16–A.21. All core exchanges are made by first lifting a core thread (yellow) from the harness and placing it on the stand in its destination position. The displaced orange thread is then moved to the harness.

Note that because Methods 1 and 2 move the threads in different sequences, the order of working the threads during a core change is also different. The middle diagram shows *the color placement of the core threads before they are moved* and the written instructions *describe the color order as it should look while the moves are being made.*

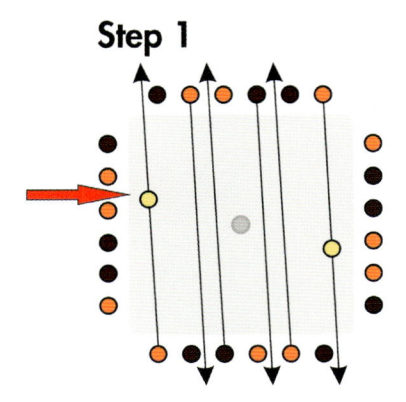

Note that in Method 1, when the braider exchanges core threads, the usual order of working pairs of threads *is changed to moving threads one thread at a time, right-to-left*. Method 2 braiders will continue to use the original method of working.

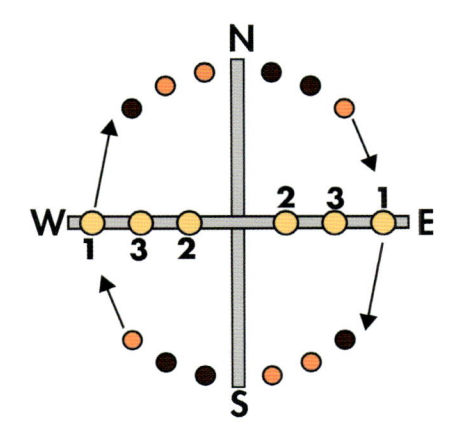

**FIGURE A.14.**
Bird's-eye view of the threads on the E/W axis of the harness and the N/S sides of the stand.

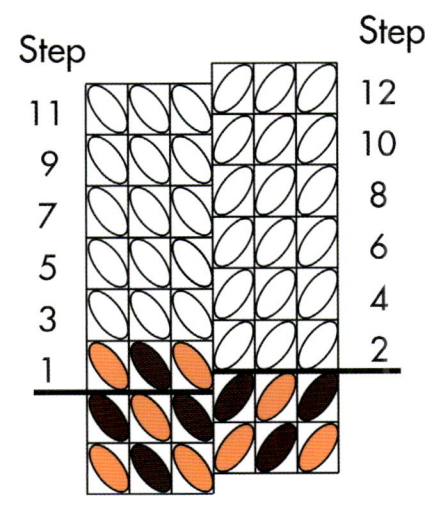

**FIGURE A.13.**
The red arrow points to the circle that indicates a core change is to be made, and the color of the core thread that will become active.

**FIGURE A.15.**
The grid shows the completed stitch and its color placement for the row started two steps previously.

*Step 1. Work the N/S core exchanges referencing figure A.16.*

**Method 1.** Starting at the right:

1. Lift the yellow #1 thread from the harness arm. Place it to the outside of the brown thread as indicated by the arrow.
2. Move the N/E orange thread to the harness.
3. Work the non-core threads one at a time from right to left. End by making a core change between the S/W orange thread and the yellow #1 core thread.

**Method 2.** Exchanging the outside threads first:

1. Lift the yellow #1 core threads from the harness arms and place them *outside* the brown threads as indicated by the arrows.
2. Move the indicated N/E and S/W orange threads to the harness.
3. Work the second and third pairs of threads as usual.

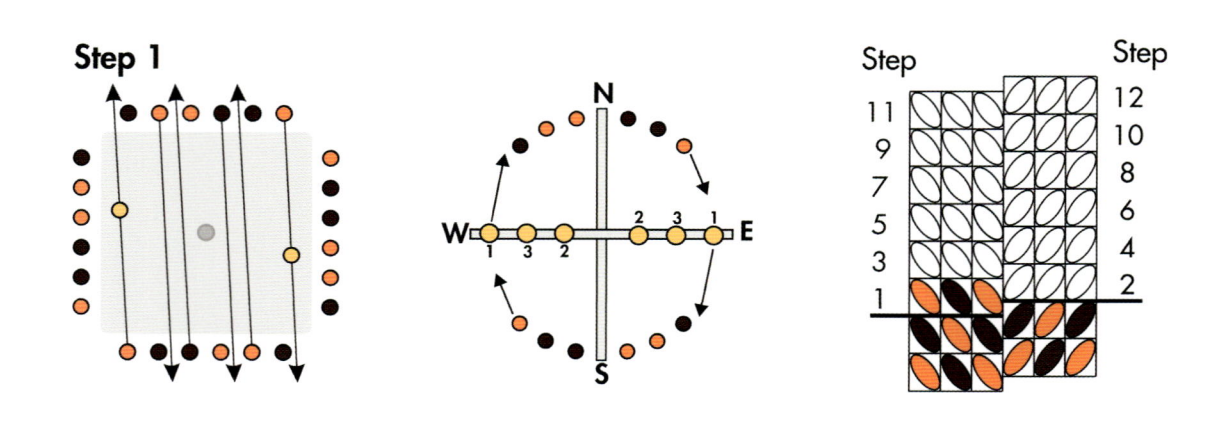

**FIGURE A.16.**
Step 1: *(Above left)* Rotation moves for N/S threads. Yellow circles indicate threads to be exchanged.
*(Above center)* Yellow #1 core threads move from the harness to the stand.
*(Above right)* Row 1 shows completed stitches started in previous moves.

*Step 2. Work the E/W core exchanges referencing figure A.17.*

**Method 1.** Stand facing the W side of the stand. Starting from the right:

1. Lift the S yellow #1 thread from the harness arm and place it outside the brown thread as indicated by the arrow.
2. Move the SW orange thread to the harness.
3. Work the non-core threads one at a time from right to left. End by making a core change between the N/W orange thread and the yellow #1 core thread.

**Method 2.**

1. Lift the yellow #1 core threads from the harness arms and place them to the *outside* of the brown threads as indicated by the arrows.
2. Move the indicated N/E and S/W orange threads to the harness.
3. Work the second and third pairs of stand threads as usual.

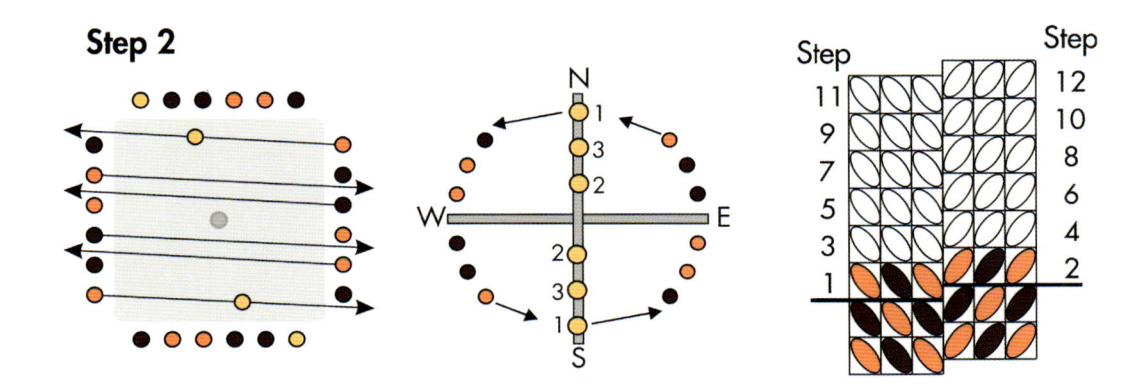

FIGURE A.17.
Step 2: *(Above left)* Rotation moves for E/W threads with first E/W core exchanges.
*(Above center)* E/W exchange of #1 core threads.
*(Above right)* Row 2 stitch and color placement.

*Step 3. Exchange the center N/S pair of threads, as seen in figure A.18.*

The braiding method used will change the order in which core exchanges are made.

**Method 1**. *Working right to left*, make the first two moves of Step 3, then:

1. Lift the yellow #2 thread from the E arm and place it on the stand to the right of the brown thread on the S side of the stand.
2. Move the N orange thread to the harness as indicated by the arrow.
3. Make the remaining core exchange between the yellow #2 W thread and the indicated orange thread on the S side of the stand.
4. Work the remaining two threads in the Step 3 diagram.

**Method 2**. Work the two outside N/S pairs in the Step diagram, then:

1. Lift the yellow #2 thread from the E arm and place it on the stand to the right of the brown thread on the S side of the stand.
2. Move the N orange thread to the harness as indicated by the arrow.
3. Make the last core exchange between the #2 W core thread and the indicated orange thread on the S side of the stand.

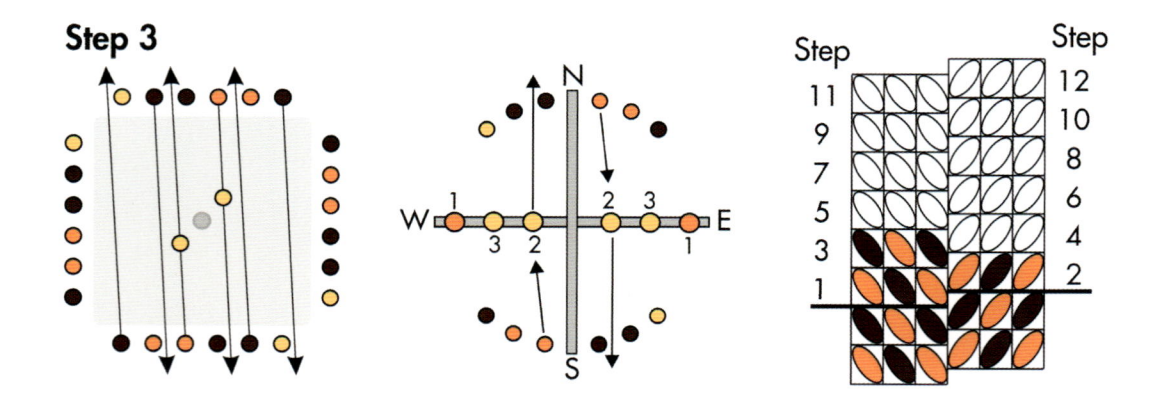

FIGURE A.18.
Step 3: *(Above left)* Rotation moves for second N/S core exchanges.
*(Above center)* The center #2 yellow core threads move from the harness down to the stand making pairs of yellow threads. The orange threads move up to the harness.
*(Above right)* Row 3 stitch and color placement.

198

*Step 4. Exchange the center E/W pair of threads referencing figure A.19.*

**Method 1.** Stand facing the W side of the equipment. Working right to left:

1. Work the first two threads as shown in the Step 4 rotation diagram.
2. Lift the yellow #2 threads from the harness arms and place them where indicated by the arrows. (The first core thread goes to the outside of the single brown thread on the stand; the second goes between two brown threads.) Move the indicated orange thread to the harness.
3. Work the remaining two non-core threads.

**Method 2.** Work the two outside E/W pairs in the Step 4 diagram, then:

1. Lift the yellow #2 core thread from the S arm and place it below the indicated brown thread on the E side of the stand.
2. Move the W orange thread to the harness.
3. Make the core exchange between the #2 N core thread and the E orange thread.

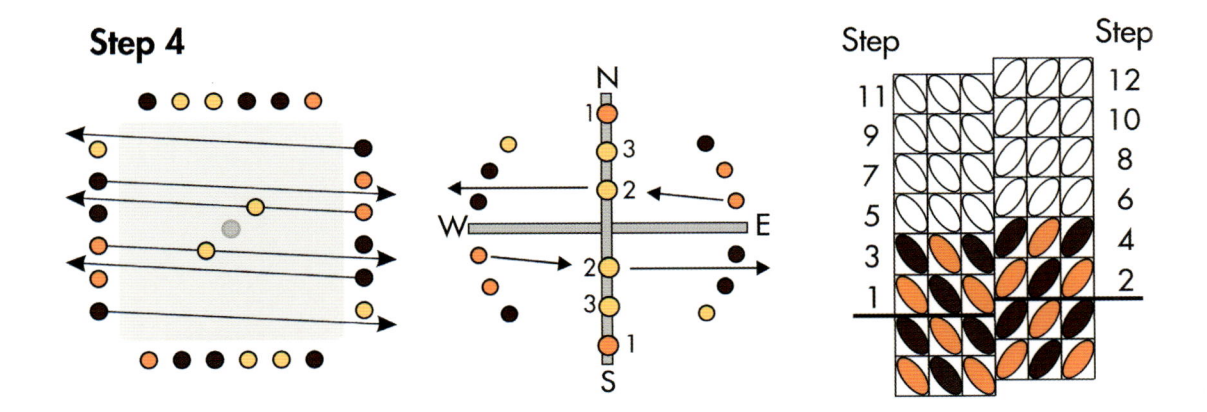

FIGURE A.19.
Step 4: *(Above left)* Rotation movements for E/W threads with the second pair of core exchanges.
*(Above center)* The yellow #2 core threads come down to the stand and are replaced by orange threads on the harness.
*(Above right)* Row 4 stitch and color placement.

*Step 5. Exchange the third pair of N/S core threads referencing figure A.20.*

The last N/S pair has been exchanged and is now in the outside position. Therefore, Step 5 becomes a repeat of Step 1, except for the harness position of the core threads. Note that the yellow threads exchanged in Step 1 are now visible both on the braid and on the stitch placement grid. Work Step 5 as follows:

**Method 1**. Starting at the right:

1. Lift the yellow #3 thread from the harness arm and place it to the outside of the brown thread as indicated by the arrow.//
2. Move the N/E orange thread to the harness.
3. Work the non-core threads, one at a time, from right to left. End by making a core change between the S/W orange thread and the yellow #1 core thread.

**Method 2**. Exchanging the outside threads first:

1. Lift the yellow #3 core threads from the harness arms and place them to the *outside* of the brown threads as indicated by the arrows.
2. Move the indicated N/E and S/W orange threads to the harness.
3. Work the second and third pairs of threads as usual.

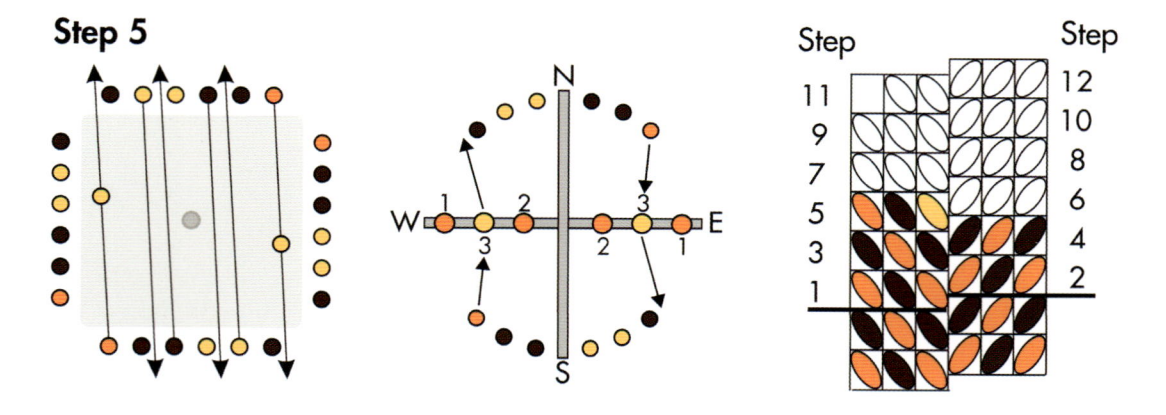

FIGURE A.20.
Step 5: Rotation movements for #3 N/S core exchanges.
*(Above center)* Yellow #3 core threads are placed in the outer positions on the stand and orange threads move to the harness.
*(Above right)* the first yellow core changes appear on the braid.

*Step 6. Exchange the third pair of E/W core threads as in figure A.21.*

Step 6 is identical to Step 2 except that the #3 core threads are exchanged instead of the #1's.

**Method 1.** Move to the W side of the stand and begin working at the right.

1. Lift the yellow #3 thread from the harness arm and place it outside of the brown thread as indicated by the arrow.
2. Move the SW orange thread to the harness.
3. Work the non-core threads, one at a time from right to left. End by making a core change between the N/W orange thread and the yellow #3 core thread.

**Method 2.**

1. Lift the yellow #3 core threads from the harness arms and place them *outside* the brown threads as indicated by the arrows.
2. Move the N/E and S/W orange threads indicated to the harness.
3. Work the second and third pairs of stand threads as usual.

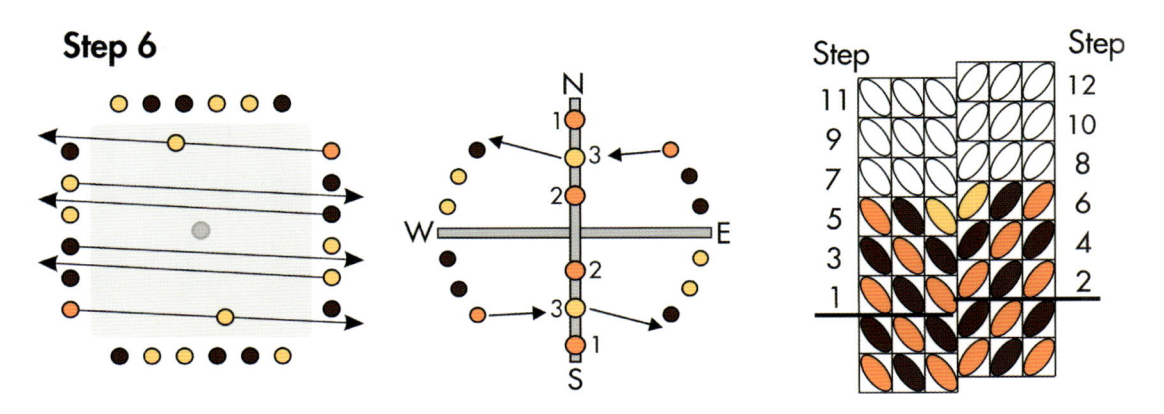

**FIGURE A.21.**
Step 6: Rotation movements for E/W threads with the yellow #3 bobbins completing the core exchanges.

---

*Steps 7 and 8. Braid the new colorway as in figure A.22.*

With the core changes complete, the new colorway is continued by repeating Steps 7 and 8. (See figure A.22.) The pattern is now yellow and brown. A variety of color effects created with core exchanges can be seen on the Design page for this pattern.

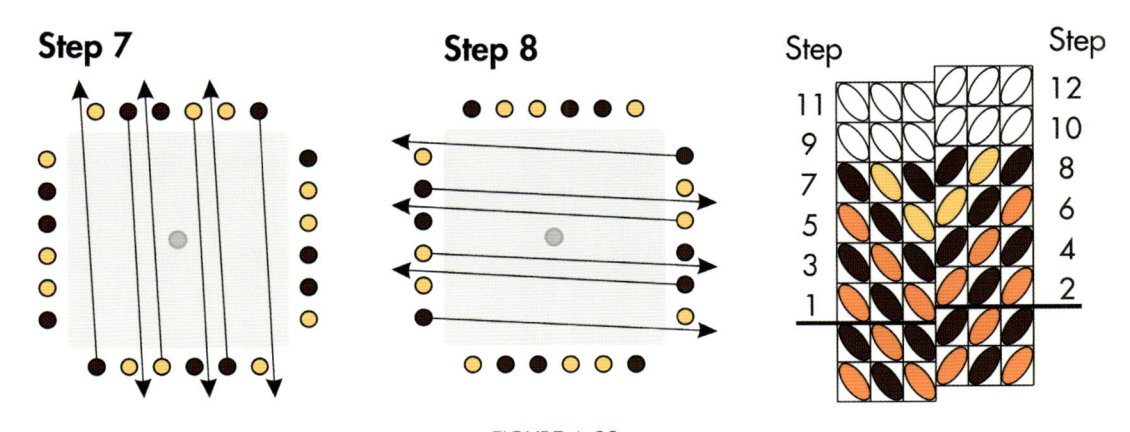

**FIGURE A.22.**
Following all core exchanges, the orange and brown braid is now yellow and brown.

# APPENDIX F
## Comparison of Sling Dimensions

| Cradle | *Style | **Image | Sling Total Length (cm) | Sling Cord 1 Strands | Sling Cord 2 Strands | ***Cradle Ribs | Cradle Length (cm) | Cradle Width (cm) | Cradle Slit (cm) |
|---|---|---|---|---|---|---|---|---|---|
|   | Macusani | Fig 5.28 | 227 | 16-24 | 24-16 | $\frac{4}{4}$ | 20 | 4.6 | 4 |
|   | Macusani | Fig 8.11 | 162 | 24 | 16 | $\frac{3\ 4\ 3}{3\ 4\ 3}$ | 10 | 3 | 2 |
|   | Macusani | Fig 1.39 | 118 | 24 | 24-16 | $\frac{8}{8}$ / $8$ | 17 | 5 | 2 |
|   | Ilave | Fig 7.1 | 195 | 24-36 | 36-24 | $\frac{4\ 6\ 4}{4\ 6\ 4}$ | 15 | 5 | 3 |
|   | Macusani | Fig 10.2 | 145 | 48 | 48 | $\frac{3\ 4\ 3}{3\ 4\ 3}$ | 10 | 3 | 1 |
|   | Ilave | Fig 1.40 | 210 | 16-24-36 | 36-24-16 | $\frac{4}{4}$ | 20 | 4 | 14 |
|   | Ilave | Fig 1.23 | 190 | 24-32 | 32-24 | $\frac{4\ 6\ 4}{4\ 6\ 4}$ | 15 | 4 | None |
|   | Ilave | Fig 1.41 | 164 | 24-36 | 36-24 | $\frac{1\ 2\ 4\ 2\ 1}{1\ 2\ 4\ 2\ 1}$ | 16 | 3 | 12 |
|   | Macusani | Fig 1.1 | 170 | 24 | 24 | $\frac{5\ 9\ 5}{4\ 7\ 4}$ | 18 | 8 | 9 |
| 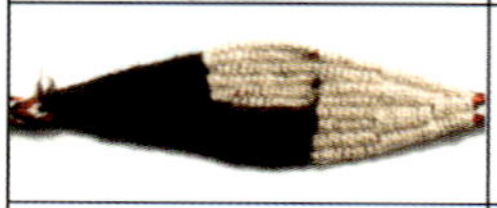  | Ilave | Fig 8.10 | 170 | 24 | 24 | $\frac{2\ 3\ 6\ 4\ 2}{2\ 3\ 6\ 4\ 2}$ | 16.5 | 5.5 | 4.5 |
|   | Ilave | Fig 10.3 | 128 | 16-24 | 24-16 | $\frac{2\ 4\ 2}{2\ 4\ 2}$ | 11 | 3.5 | 8 |

*Area Style: based on Elayne Zorn's paper, "Sling Braiding in the Mascusani Area of Peru." Textile Musuem Journal, (1980-1981): 41-54.
**Image: Shows the figure number where the complete sling or partial sling can be seen.
***Cradle Ribs: Shows how the ribs of the cradle were divided when woven, reading from left to right.

FIGURE A.23.
Comparison of the dimensions of a selection of contemporary slings. *Collection of R. Owen.*

# NOTES

## Chapter 1. A History of the Sling

1. Ferrill, *Origins of War*, 24-26.
2. Harrison, "Sling in Medieval Europe," 34-35.
3. Korfman, "Sling as a Weapon," 42.
4. Greep, "Lead Slingshot from Windrush Farm, St Albans," 184.
5. Kelly, "The Cretan Slinger," 273.
6. Korfman, "Sling as a Weapon," 39.
7. Smith, *Dictionary of Greek*, 554.
8. Ibid.
9. Strabo, *Geography of Strabo*, 168.
10. Richardson, "Ballistics of the Sling," 44-49.
11. Vegetius Renatus, *Military Institutions*, 75.
12. Richmond, "Trajan's Army," 14-17.
13. Finney, "Middle Iron Age Warfare," 9.
14. York, *Slings and Slingstones*, 56.
15. Strutt, *Sports and Pastimes*, 60.
16. Ibid.
17. Morillo, *Battle of Hasting*, 31.
18. Dunsmore, *Nepalese Textiles*, 170-171.
19. Ibid.
20. d'Harcourt, Raoul. *Textiles of Ancient Peru*, 110.
21. Christian, F.W., "On Micronesian Weapons," 297.
22. Te Rangi Hiroa, "Cook Islands," 300-301.
23. Lèvesque, *History of Micronesia*, 39.
24. York, *Slings and Slingstones*, 23.
25. Stov, "Experimentation in Sling Weaponry," 15.
26. Heizer, "Prehistoric Sling from Lovelock," 139.
27. Hill, *Agricultural and Hunting Methods*, 38.
28. Stov, "Experimentation in Sling Weaponry," 116.
29. York, 89.
30. Ibid., 91.
31. Akush, "War, Chronology, and Causality," 4.
32. Contreras, "Huaqueros and Remote Sensing Imagery," 545.
33. Proulx, "Ritual Uses of Trophy Heads," 125.
34. Atwood, *Lives in Looting*, n.p.
35. Bird, "Preceramic Excavations," 101.
36. Ibid., 212.
37. Ibid., 214.
38. Reiss, *The Necropolis of Ancon in Perú*, 99.
39. Bird, "Preceramic Excavations," 6.
40. Ibid., 112-117, 168.
41. Engel, "Preceramic Settlement," 57.
42. Dwyer, "The Chronology and Iconography," 105-128.
43. Paul, *Paracas Ritual Attire*, 4.
44. Peters, "Identity, Innovation, and Textile Exchange," 1.
45. O'Neale, "Textiles of Early Nazca Period," 127-128.
46. Paul, *Paracas Ritual Attire*, 5.
47. Ibid., 14.
48. Ibid., 39.
49. Ibid., 137.
50. O'Neale, "Textiles of Early Nazca Period," 201, plates XLVIII a-c.
51. Ibid.
52. Ibid., plate XLVIII d.
53. d'Harcourt, *Textiles of Ancient Peru*, 89.
54. O'Neale, "Textiles of Early Nazca Period," 202.
55. d'Harcourt, 128.
56. Proulx, "Ritual Uses of Trophy Heads," 127.
57. "Archaeology in Ancon," n.p.
58. Slovak, "Reconstructing Middle Horizon Mobility," 157-165.
59. Reiss, *The Necropolis of Ancón in Perú*, 99.
60. Quilter, *Art and Moche Material Arts*, 212.
61. Ibid., 222.
62. Zorn, "Transformation in Tapestry," 478.
63. Quilter, *The Ancient Central Andes*, 200.
64. Tung, "Trauma and Violence," 241.
65. Quilter, *The Ancient Central Andes*, 262.
66. Ibid., 263.
67. Rowe, "Inca Culture," 113.
68. Ibid., 143.
69. Cobo, *Inca Religion and Customs*, 32.
70. Rowe, "Inca Culture," 295.
71. Ibid., 279.
72. Filipe Guaman Poma de Ayala, *The First New Chronicle*, 109.

73  Cobo, *Inca Religion and Customs*, 215.
74  Ibid.
75  Ibid., 128.
76  Rowe, "Inca Culture," 278.
77  Ibid., 275.
78  Malpass, *Daily Life in the Inca Empire*, 77.
79  Frye, *First New Chronicle*, 275.
80  Rowe, "Inca Culture," 217.

## Chapter 3. Core Braiding Yarns: History, Contemporary Choices, and Design Considerations

1  https://en.wikipedia.org/wiki/Camelid, May 29, 2016.
2  Bird, "Fibers and Spinning Procedures," 15.
3  Lichtenstein, "Vicugna, Vicugna," n.p.
4  Bird, "Fibers and Spinning Procedures," 16.
5  Wheeler, "Pre-Conquest Alpaca and Llama Breeding," n.p.
6  Ibid.
7  Ibid.
8  O'Neale, "Textiles of Early Nazca Period," 202.
9  Crawford, "Peruvian Textiles," 64.
10  O'Neale, 145.
11  Ibid., 145.
12  Alvarez, *Textile Traditions of Chinchero*, 76–77.
13  Zorn, "Sling Braiding in the Macusani Area," 48-49.
14  Alvarez, *Andean Spinning* (video).
15  Cahlander, *Sling Braids of the Andes*, 30.
16  d'Harcourt, *Textiles of Ancient Peru*, 93–95.
17  Alvarez, *Andean Spinning* (video).
18  Ibid.
19  Ibid.
20  Alvarez, *Weaving in the Peruvian Highlands*, 48.
21  Robson, *The Fleece and Fiber Sourcebook*, 11.
22  Porter, Alderson, Hall, and Sponenberg, *Mason's World Encyclopedia of Livestock Breeds and Breeding, 2 Volume Pack*, 731.
23  Terriss, Bernardo Fulcrand, *Improving the Performance of Indigenous Sheep Breeds*, 1–2.
24  http://www.britishwool.org.uk/assets/uploads/sheep_breeds/Leicester%20Longwool%20LUSTRE.pdf.
25  Wheeler, "Pre-Conquest Alpaca and Llama Breeding," n.p.
26  "About Suri Alpacas."
27  Robson, *The Fleece and Fiber Sourcebook*, 365.
28  Alvarez, *Andean Spinning* (video).
29  Means, "A Series of Ancient Andean Textiles."

## Chapter 6. Beginnings, Endings, and Embellishments

1  Owen, *Braids, 250 Patterns*, 43.
2  Owen, Flynn, *Andean Sling Braids, New Techniques for Artists*, 26–27.
3  O'Neale, "Textiles of Early Nazca Period," 177.

## Chapter 7. Making an Andean-style Sling

1  Zorn, "Sling Braiding in the Mancusani Area of Peru," 43-45.
2  Cahlander, *Sling Braids of the Andes*, 31.
3  Zorn, "Sling Braiding in the Macusani Area of Peru," 43.
4  Cahlander, *Sling Braids of the Andes*, 31.
5  Ibid., 30.
6  Ibid., 30.

# BIBLIOGRAPY

"About Suri Alpacas." Surinetwork (website). http://www.surinetwork.org/AboutSuriAlpacas.

Akush, Elizabeth. "War, Chronology, and Causality in the Titicaca Basin." *Latin American Antiquity* 19, no. 4 (2008): 4.

Alvarez, Nilda Callañaupa. *Andean Spinning with Nilda Callañaupa Alvarez*. Loveland, Colo: Interweave Video Studios, 2014.

———. *Textile Traditions of Chinchero: A Living Heritage*. Loveland, Colo: Thrums LLC, 2012.

———. *Weaving in the Peruvian Highlands: Dreaming Patterns, Weaving Memories*. Cusco, Peru: Center de Textiles Traditionales del Cusco, 2007.

"American Section: Mesoamerica." http://www.penn.museum/about-our-collections/american-section.html.

"Archaeology in Ancon." 2009. http://www.museodeancon.com/ingles/arqueologiaenancon.php.

Atwood, Roger. *Lives in Looting: How Professional Graverobbers Are Destroying the Past*. Washington, DC: Alicia Patterson Foundation, 2004. http://aliciapatterson.org/stories/lives-looting-how-professional-grave-robbers-are-destroying-past.

"Augustus Pitt-Rivers." http://en.wikipedia.org/wiki/Augustus_Pitt-Rivers.

Bankes, George. *Peru before Pizarro*. Oxford: Phaidon Press Ltd., 1977.

Bird, Junius. "Fibers and Spinning Procedures in the Andean Areas," The Junius B. Bird Pre-Columbian Textile Conference, May 19th and 20th, 1973, Washington, D.C. *Proceedings* edited by Ann P. Rowe, et al. 16.

———. "The Preceramic Excavations at the Huaca Prieta, Chicama Valley, Peru." *American Museum of Natural History. Anthropological Papers* 62, part 1 (1985): 214. http://hdl.handle.net/2246/241.

Bird, Junius, and Louisa Bellinger. *Catalogue raisonne [of the Textile Museum]. Paracas Fabrics and Nazca Needlework: 3rd century B.C.–3rd century A.D.* Washington, D.C.: Textile Museum, 1954.

Browman, David L. review of *The Life and Writings of Julio C. Tello*, edited by Richard L. Burger (2009), in *Bulletin of the History of Archaeology* 20, no. 2 (2010). http://www.archaeologybulletin.org/rt/printerFriendly/bha.20205/25.

Brown Vega, Margaret, and Nathan Craig. "New Experimental Data on the Distance of Sling Projectiles." *Journal of Archaeological Science* 36 (2009): 1264-1268. http://www.academia.edu/176644/New_Experimental_Data_on_the_Distance_of_Sling_Projectiles.

Bruhn de Hoffmeyer, Ada. *Arms and Armour in Spain, a Short Survey, II*. Madrid: Instituto de Estudios sobre Armas Antiguas, CSIC, 1982.

Cahlander, Adele, Elayne Zorn, and Ann Pollard Rowe. *Sling Braiding of the Andes*. Boulder: Colorado Fiber Center, 1980.

Christian, F. W. "On Micronesian Weapons, Dress, Implements, etc." *Journal of the Anthropological Institute of Great Britain and Ireland* 28, no. 3/4 (1899): 288-306. http://www.jstor.org/discover/10.2307/2842880?uid=3738032&uid=4577293337&uid=2&uid=3&uid=60&sid=21102628386043.

Cobo, Bernabé. *Inca Religion and Customs*. Translated by Roland Hamilton. Austin: The University of Texas Press, 1990.

Contreras, Daniel A. "Huaqueros and Remote Sensing Imagery: Assessing Looting Damage in the Viru Valley, Peru." *Antiquity* 84, no. 324 (2009): 544-555. http://antiquity.ac.uk/ant/084/ant0840544.htm.

Crawford, M.D.C. "Peruvian Textiles." *American Museum of Natural History, Anthropological Papers,* 12, pt. 3. (1915): 64, 153.

Daggett, Richard E., "Julio C. Tello: An Account of His Rise to Prominence in Peruvian Archaeology." In *The Life and Writings of Julio C. Tello: America's First Indigenous Archaeologist*, edited by Richard L. Burger. Iowa City: University of Iowa Press, 2009.

Dohrenwend, Robert E. "The Sling: Forgotten Firepower of Antiquity." *Journal of Asian Martial Arts* 11, no 2 (2002): 31-32. http://www.journalofasianmartialarts.com/product/asia/weaponry/the-sling-forgotten-firepower-of-antiquity-detail-255.

Dunsmore, Suzi. *Nepalese Textiles*. London: British Museum Press, 1993.

Dwyer, Jane P. "The Chronology and Iconography of Paracas-style Textiles," The Junius B. Bird Pre-Columbian Textile Conference, May 19th and 20th, 1973, Washington, D.C. *Proceedings*, edited by Ann P. Rowe, et al. 105-28.

Echols, Edward C. "The Ancient Slinger." *The Classical Weekly* 43, no. 15 (1950): 227-230. http://www.jstor.org/stable/4342755.

Engel, Frederic A. "A Pre-ceramic Settlement on the Central Coast of Peru: Asia, Unit 1." *Transactions of the American Philosophical Society*, 53, no. 3 (1963): 57.

Ferrill, Arther. *The Origins of War: The Stone Age to Alexander the Great*. London: Thames & Hudson, 1985. 24-26.

Finney, J. B. "Middle Iron Age Warfare of the Hillfort Dominated Zone, c.400 B.C. to c.150 B.C." PhD thesis, Bournmouth University, 2005. *BAR British Series* 423 (2006): 9. http://eprints.bournemouth.ac.uk/396/.

Fleming, Stuart. "The Mummies of Pachacamac – An Exceptional Legacy from Uhle's 1896 Excavations." *Expedition* 28, no. 3 (Winter 1986): 39.

Foss, Clive. "Greek Sling Bullets in Oxford." *The Society for the Promotion of Hellenic Studies. Archaeological Reports* 21 (1974-75): 40-44. http://www.jstor.org/discover/10.2307/581133?uid=3738032&uid=4577293337&uid=2129&uid=2&uid=70&uid=3&uid=60&sid=21103605671547.

Frye, David. *The First New Chronicle and Good Government,* (Abridged) Selected, translated and annotated by Felipe Guaman Poma de Ayala. Cambridge: Hackett Publishing Co., Inc., 2006.

Gaustad, Stephenie. *The Practical Spinner's Guide: Cotton, Flax, Hemp.* Loveland, Colo: Interweave Press, 2014. 53–62.

Greep, S. J. "Lead Sling-shot from Windridge Farm, St. Albans and the Use of the Sling by the Roman Army in Britain." *Britannia* 18 (1987): 183-200. http://www.jstor.org/discover/10.2307/526445?uid=3738032&uid=2129&uid=2134&uid=2&uid=70&uid=4&uid=21103509561113.

Guaman Poma de Ayala, Filipe. *The First New Chronicle and Good Government on the History of the World and the Incas up to 1615.* Translated and edited by Roland Hamilton. Austin: University of Texas Press, 2009.

d'Harcourt, Raoul. *Textiles of Ancient Peru and Their Techniques.* Edited by Grace G. Denny and Carolyn M. Osborne. Seattle: University of Washington Press, 1962.

Harrison, Chris. "The Sling in Medieval Europe." *The Bulletin of Primitive Technology* 31 (Spring 2006). http://www.chrisharrison.net/index.php/Research/Sling.

Hawkins, Walter. "Observations on the Use of the Sling, as a Warlike Weapon, Among the Ancients." *Publisher Society of Antiquaries of London.* January 1, 1847. https://play.google.com/books/reader?id=CTEUAAAAYAAJ&printsec=frontcover&output=reader&authuser=0&hl=en&pg=GBS.PA98.

Heizer, R. F., and I. W. Johnson. "A Prehistoric Sling from Lovelock Cave, Nevada.*" American Antiquity* 18, no. 2 (October 1962): 139-147.

Hicks, Dan, and Alice Stevenson, eds. *World Archaeology at the Pitt Rivers Museum: A Characterization.* Oxford: Archaeopress, 2013.

Hill, Willard Williams. *The Agricultural and Hunting Methods of the Navaho Indians.* New Haven, CT: Yale University Press; London: Oxford University Press, 1938.

Hogg, Oliver. *Clubs to Cannons.* London: Duckworth & Co. Ltd., 1968.

"Inca Army." http://en.wikipedia.org/wiki/Inca_army.

Karoll, Amy. "A Comparative Study of the Swennes Woven Nettle Bag and Weaving Techniques." *Journal of Undergraduate Research XII* (2009). http://www.uwlax.edu/urc/jur-online/PDF/2009/karoll-amyARC.pdf.

Kelly, Amanda. "The Cretan Slinger at War – A Weighty Exchange.*" Annual of the British School of Athens* 107 (2012): 273-311. http://dx.doi.org/10.1017/S006824541200007X.

Korfman, Manfred. *"*The Sling as a Weapon.*" Scientific American* 229, no 4 (October 1973): 35, 39, 42.

Kroeber, Alfred. "Archaeological Explorations in Peru: Part 1, Ancient Pottery from Trujillo." *Field Museum of Natural History [Anthropology, Memoirs]* 2, no. 1 (1926). https://archive.org/details/archaeologicalex21kroe.

Lambert, Patricia, Barbara Staepelaere, and Mary G. Fry. *Color and Fiber.* Atglen, PA: Schiffer Publishing, Ltd., 1986.

Léry, Jean de. *The Memorable History of the Siege of Sancerre* (1574). http://en.wikipedia.org/wiki/Siege_of_Sancerre.

Lèvesque, Rodrigue. *History of Micronesia: Focus on the Mariana Mission, 1670-1692.* Gatineau, Quebec: Editions Levesque, 1995.

Lichtenstein, G. R., et al. "Vicugna vicugna." *The IUCN Red List of Threatened Species – 2014.3.* http://www.iucnredlist.org/details/22956/0.

MacQuarrie, Kim. *The Last Days of the Incas.* New York: Simon & Schuster, 2007.

Malpass, Michael A. *Daily Life in the Inca Empire.* Westport, Conn: Greenwood Press, 1996.

Markham, Clements R. *The Second Part Chronicle of Peru of Pedro de Cieza de Leon.* London: Hakluyt Society Whiting and Co., 1883. https://archive.org/details/secondpartchron00markgoog.

Martin, Luis. *The Kingdom of the Sun: A Short History of Peru.* New York: Charles Scriber's Sons, 1974.

Mason, Alden J. *The Ancient Civilizations of Peru.* New York: Penguin Books. 1978.

McEwan, Gordon Francis. *The Incas: New Perspectives.* New York: W.W. Norton, 2008.

Means, Philip Ainsworth. "A Series of Ancient Andean Textiles." *Bulletin of the Needle and Bobbin Club* 11, no. 1 (1927).

Morillo, Stephen, ed. *The Battle of Hastings: Sources and Interpretations.* Woodbridge, UK: Boydell & Brewer, 1996. [book review] http://books.google.com.au/books?id=4jvgO9opigC&q=slingers#v=snippet&q=slingers&f=false.

Natural Colored Wool Growers Association. (website) http://www.ncwga.org/.

Noble, Carol Rasmussen. "Peruvian Slings: Their Uses and Regional Variations." *The Weaver's Journal,* VI, no. 4, (1982): 53-56. http://www.cs.arizona.edu/patterns/weaving/periodicals/wj_24.pdf.

"Of Alcazar Fortress in Toledo," YouTube video, 8:15. http://www.youtube.com/watch?v=eksEL8qhP-w.

Onasander. *The General (Strategikos).* Harvard University Press. Loeb Classical Library edition, 1928. http://penelope.uchicago.edu/Thayer/E/Roman/Texts/Onasander/H*.html.

O'Neale, Lila. "Textiles of Early Nazca Period." *Anthropological Memoirs*, II, no 3 [Field Museum of Natural History, Chicago] (1937). 127-128. http://www.jstor.org/stable/275691?origin=JSTOR-pdf.

Orlove, Benjamin S. *Alpacas, Sheep, and Men: The Wool Export Economy and Regional Society in Southern Peru.* New York: Academic Press, 1977.

Owen, Rodrick. *Braids, 250 Patterns from Japan, Peru and Beyond.* Loveland, Colo.: Interweave Press, 1995.

Owen, Rodrick, and Terry N. Flynn. *Andean Sling Braids: New Designs for Textile Artists.* Atglen, PA: Schiffer Publishing, Ltd., 2016.

Paul, Anne. *Paracas Ritual Attire, Symbols of Authority in Ancient Peru.* Norman: University of Oklahoma Press, 1990.

Peters, Ann H. "Identity, Innovation, and Textile Exchange Practices at the Paracas Necropolis: 2000 BP," *Textile Society of America Symposium Proceedings* Paper 726 (2012): 1. http://digitalcommons.unl.edu/cgi/viewcontent.cgi?article=1725&context=tsaconf.

Pritchett, William Kendrick. *The Greek State at War.* Berkeley: University of California Press, 1974.

Proulx, Donald. A. "Ritual Uses of Trophy Heads in Ancient Nazca Society." In *Ritual Sacrifice in Ancient Peru,* ed. by Elizabeth Benson and Anita Cook, 119-136. Austin: University of Texas Press, 2001. http://people.umass.edu/proulx/online_pubs/Ritual_Uses_of_Trophy_Heads_Texas.pdf.

———. *A Sourcebook of Nasca Cermain Iconography: Reading a Culture Through Its Art.* Iowa City: University of Iowa Press, 2009.

Quilter, Jeffrey. *The Ancient Central Andes.* London: Routledge, 2014.

———. *Art and Moche Material Arts.* Edited by Steve Bourget and Kimberly L. Jones. Austin: University of Texas Press, 2008.

Ramirez, Susan E. *To Feed and Be Fed: The Cosmological Bases of Authority and Identity in the Andes.* Stanford, CA: Stanford University Press, 2005.

Reiss, W., and Alphons Stubel. *The Necropolis of Ancon in Perú.* Berlin: A. Asher & Co., 1880-1887. http://pds.lib.harvard.edu/pds/view/13890340?n=113&imagesize=1200&jp2Res=0.125&printThumbnails=no.

Richardson, Thom. "The Ballistics of the Sling." *Royal Armouries Yearbook 3.* (1998):44-49. http://slinging.org/index.php?page=the-ballistics-of-the-sling---thom-richardson.

Richmond, I. A. "Trajan's Army on Trajan's Column." *Papers of the British School at Rome.* 1935. http://www.jstor.org/discover/10.2307/40310440?uid=3738032&uid=2&uid=4&sid=21104526665763.

Robson, Deborah, and Carol Ekarius. *The Fleece and Fiber Sourcebook: More Than 200 Fibers, from Animal to Spun Yarn.* North Adams, Mass: Storey Publishing, 2011.

Rowe, Howland. J. "Inca Culture at the Time of the Spanish Conquest." In *Handbook of South American Indians,* 2. Julian H. Steward, ed. Washington, DC: Smithsonian Institution. 1946. www.lib.berkeley.edu/ANTH/emeritus/rowe/pub/rowe.pdf.

"Sling Ranges [analysis of projectile weight and distance thrown]," http://slinging.org/index.php?page=sling-ranges.

Slovak, Nicole, et al. "Reconstructing Middle Horizon Mobility Patterns on the Coast of Peru Through Strontium Isotope Analysis." *Journal of Archaeological Science* 36 (2009): 157-165.

Smith, William. *A Dictionary of Greek and Roman Antiquities.* London: John Murray, 1875. Retrieved from *Greek Roman Dictionary.* http://penelope.uchicago.edu/Thayer/E/Roman/Texts/secondary/SMIGRA*/Funda.html.

Speiser, Noemi. *The Manual of Braiding.* Basel, Switzerland: N. Speiser, 1983.

"Stages of Occupation at Ancon." http://www.museodeancon.com/ingles/arqueologiaenancon.php.

Stov, Eric T. *Experimentation in Sling Weaponry: Effectiveness of, and Archaeological Implications for a World-Wide Primitive Technology.* MA Thesis, Dept. of Anthropology, University of Nebraska. 2013. http://digitalcommons.unl.edu/anthrotheses/30/.

Strabo. *The Geography of Strabo,* 2, Book III, Chpt.5, p. 168. Harvard University Press. Loeb Classical Library edition, 1923. http://penelope.uchicago.edu/Thayer/E/Roman/Texts/Strabo/3E*.html.

Strutt, Joseph, and J. Charles Cox. "Rural Exercises Generally Practised." *The Sports and Pastimes of the People of England.* (2nd ed.) London: Methuen & Co., 1903. http://www.sacred-texts.com/neu/eng/spe.

"Synthetic Fibers." Technology Transfer Network. United States Enviromental Protection Agency, September 1990. http://www.epa.gov/ttnchie1/ap42/ch06/final/c06s09.pdf.

Tada, Makiko. *Comprehensive Treatise of Braids II, Andean Sling Braids.* Tokyo: Text, Inc., 1996.

Te Rangi Hiroa (Peter H. Buck). "Arts and Crafts of the Cook Islands: Weapons," *Bernice P. Bishop Museum Bulletin* 179 (1944). http://nzetc.victoria.ac.nz/tm/scholarly/tei-BucArts-t1-body-d1-d15.html#n307.

Tung, Tiffiny A. "Truama and Violence in the Wari Empire of the Peruvian Andes: Warfare, Raids, and Ritual Fights." *American Journal of Physical Anthropology* 133, no. 3 (July 2007): 241.

Urton, Gary. "Moieties and Ceremonialism in the Andes: The Ritual Battles of the Carnival Season in Southern Peru," SENRI *Ethnological Studies* 37 (1993): 117-139. http://camel.minpaku.ac.jp/dspace/bitstream/10502/722/1/SES37_007.pdf.

Vegetius Renatus, Flavius. *The Military Institutions of the Romans.* Thomas R. Phillips, editor; John Clark, translator. Montana: Kessinger Publishing, 2010.

Von Hagen, Victor Wolfgang. *The Ancient Sun Kingdoms of the Americas.* Granada: Publishing Ltd., 1977.

Wheeler, Jane C. "Pre-Conquest Alpaca and Llama Breeding," *The Camelid Quarterly*, December 2005. http://www.conopa.org/publicaciones/preconquest_alpaca_and_llama_breeding.php.

York, Robert, and Gigi York. *Slings and Slingstones: the Forgotten Weapons of Oceania and the Americas.* Rev. ed. The Kent State University Press, 2011.

Zorn, Elayne. "Sling Braiding in the Macusani Area of Peru." *Textile Museum Journal,* (1980-1981): 41-54.

———. "Transformation in Tapestry in the Ayacucho Region of Peru." *The Textile Society of America Symposium Proceedings* (2004): 478.

**Rodrick Owen's** more than four decades of research into the braided textiles of pre-Hispanic Peru has helped raise awareness and appreciation for these gems of textile art. He trained at the London College of Furniture, completing the Creative Textiles Programme and qualifying with distinctions. An internationally respected braiding teacher, he popularized kumihimo, introducing takadai braiding to the US in 1995. His previous books include *Braids: 250 Patterns from Japan, Peru and Beyond* and *Making Kumihimo: Japanese Interlaced Braids*. Owen has exhibited and taught in the United Kingdom, United States, Japan, Europe, and Australia. His current research involves braided headband structures from pre-Hispanic Peru. Owen lives in Oxford, England.

**Terry Newhouse Flynn** is a textile artist and art educator who creates garments and sculptures that combine her interests in woven, braided, and knitted textiles. She has been weaving for more than thirty-five years and teaching textile arts for three decades. She credits ten years of daily immersion in running a weaving/braiding/knitting shop for deepening her understanding of yarns, color interaction, textile structure, and clothing design. She has worked with Rodrick Owen since 1993, contributing shadow weave and hemstitching to takadai braiding's repertoire. She holds a BFA in fiber and an MAT in art education from Maryland Institute College of Art. She teaches art and textile arts in Baltimore, Maryland.